PRAISE FOR THE AUT

T0009633

Leadership Moments from NASA
— Dave Williams and Elizabeth Howell

"This is a fascinating read, extracting leadership lessons from many people who were on the front lines at NASA. There is a lot of history here, and from that, one hopes, some guidance for the future."

— Mike Griffin, former NASA Administrator and
Under Secretary of Defense for Research and Engineering

"Spaceflight is a demanding and unforgiving environment and flying humans in space takes tremendous leadership skills to be successful. This book is a must-have resource and guide for anyone studying and wanting to improve their leadership skills, even in fields outside of human spaceflight."

— Bill Gerstenmaier, Vice President of Mission Assurance at SpaceX

Canadarm and Collaboration
— Elizabeth Howell, foreword by Dave Williams

"Be sure to read *Canadarm and Collaboration* for a fascinating look at Canada's evolving space program and its past, present and possible future."

— *Universe Today*

"Illustrates how the country has maintained a human spaceflight program for several decades through a combination of technological specialization — Canadarm and its successors — and collaboration with the United States."

— *The Space Review*

Defying Limits
— Dave Williams

"Williams is at his best when describing astronaut training, from the high-altitude chamber meant to help would-be astronauts recognize . . . oxygen deficiency, to . . . an aircraft fondly known as the 'vomit comet.' Space may be where astronauts 'defy limits,' but Williams's memoir reveals an astronaut's most important work takes place with feet firmly on the ground."

— *Washington Post*

"An inspirational tale of a remarkable Canadian doctor, astronaut, space walker, aquanaut, CEO and loving father who turned failure into astounding accomplishments in space and on the ground. A fabulous example of how to live life to the fullest."

— Bob McDonald, CBC's chief science correspondent and
host of *Quirks & Quarks*

WHY AM I
TALLER?

What Happens to an Astronaut's Body in Space

WHY AM I TALLER?

Dr. **DAVE WILLIAMS**, Astronaut

AND **ELIZABETH HOWELL**, PhD

LIBRARY AND ARCHIVES CANADA CATALOGUING
IN PUBLICATION

Title: Why am I taller? : what happens to an astronaut's body in space / Dr. Dave Williams, astronaut, and Elizabeth Howell, PhD

Names: Williams, Dave (Dafydd Rhys), 1954- author. | Howell, Elizabeth, 1983- author.

Identifiers: Canadiana (print) 20220232431 | Canadiana (ebook) 2022023244X

ISBN 978-1-77041-596-6 (softcover)
ISBN 978-1-77090-549-8 (ePub)
ISBN 978-1-77090-739-3 (PDF)
ISBN 978-1-77090-740-9 (Kindle)

Subjects: LCSH: Space flight—Physiological aspects. | LCSH: Space flight—Physiological effect. | LCSH: Manned space flight. | LCSH: Human physiology. | LCSH: Space medicine.

Classification: LCC RC1150 .W55 2022 | DDC 612/.0145—dc23

Copyright © Dr. Dave Williams and Elizabeth Howell, 2022

Published by ECW Press
665 Gerrard Street East
Toronto, Ontario, Canada M4M 1Y2
416-694-3348 / info@ecwpress.com

Cover design: David A. Gee
Astronaut image: iStock.com/Paul Campbell

This book is funded in part by the Government of Canada. *Ce livre est financé en partie par le gouvernement du Canada.* We also acknowledge the support of the Government of Ontario through the Ontario Book Publishing Tax Credit, and through Ontario Creates.

In memory of
Charles A. Berry, MD
"The Astronauts' Doctor"

TABLE OF CONTENTS

PREFACE

"Earth is the cradle of humanity, but one cannot live in a cradle forever."

— KONSTANTIN TSIOLKOVSKY

As the golden rays of the slowly setting sun emerge over the horizon, the hint of darkness grows. The twinkling bright light of Venus that has captured the imagination of stargazers throughout history appears in the ever-deepening dark blue above. Commonly referred to as the evening star, it is the prelude to the wonder of the night sky and the many constellations that fascinate us, just as they did the early astronomers. In far northern or southern regions, some may be lucky enough to see the magic of the aurora dancing in the heavens. This spectacular ever-changing vista of lights was thought by some ancient societies to represent the forces of good and evil dragons, their fire battling on high. Others felt that the shifting curtain of lights represented lost loved ones trying to communicate with friends and relatives back on Earth. The sense of awe and wonder invoked by the darkness above has touched onlookers' spirits for millennia, a reminder of the fragility of our shared human existence in the vast infinite void of space. Throughout history, looking up at the night sky has inspired deep curiosity about what may be out there. Those feelings were a major force that indelibly shaped my future.

Growing up in what was one of the most remarkable decades of exploration in history, I was a typical child of the '60s. Virtually every waking hour was spent outdoors, especially in the summertime. My earliest recollections go back to when I was five or six years old, when my friends and I would lie on our backs enveloped by the smell of fresh grass and gaze upward, mesmerized by the stars above, challenging each other to identify the few constellations we knew. These were the early days of space exploration and there were few human-made satellites to be seen. When we spotted them as small, moving, faint lights, our imagination immediately made us wonder if they were UFOs, alien spacecraft coming to visit our planet — a popular topic of comic books at that time. Little did I realize then that my childhood dreams of exploring space and the undersea world would one day come true.

Given what appeared to be an impossible path to become an astronaut, exploring the oceans seemed a more achievable goal and I was fortunate to learn to scuba dive when I was 12. *The Undersea World of Jacques Cousteau* was a popular TV series at the time and every week I vicariously participated in the exploits of Cousteau's team aboard the *Calypso*. Over time, my passion for diving grew into a broader desire to understand how the human body adapted to living underwater in undersea habitats. Similarly, my interest in spaceflight, inspired by the NASA missions of the Mercury, Gemini and Apollo astronauts, made me wonder about the remarkable ability of humans to thrive in such different worlds — the frontiers of space and the ocean. That quest for knowledge took me to McGill University on a 12-year journey that included studying comparative physiology and neuroscience, as an undergraduate and in graduate school, then going on to medical school.

When I responded to the Canadian Space Agency's call in 1992 for applicants to the second group of astronauts to be hired in Canada, I was working as Director of the Department of Emergency Services at Sunnybrook Regional Health Centre in Toronto. Emergency medicine specialty training had become recognized throughout North America in the preceding decade, and I was proud to have finished residency training in family medicine and specialty training in emergency medicine and then join a team of experienced clinicians able to deal with any medical or surgical emergency. Many of my colleagues and I were trauma team leaders. As well, our group provided the base hospital support for land paramedics and air ambulance attendants transporting critically ill patients. There was very little I had not seen in that role, at least on Earth. The experience was invaluable in becoming a physician astronaut (or astronaut physician — the perspective varies but with the goal of furthering the field of space medicine I tend to use physician astronaut).

In the first 60 years of human spaceflight fewer than 600 people have travelled to space, roughly the number of people that might live in a small village. They have spent 161 person-years living and working in low Earth orbit and on the lunar surface, testing the limits of human performance in exploring the extreme, harsh environment of space. Less than 50 of those astronauts have been physicians and I felt fortunate to become part of that group. Despite the great honor, and the excitement I felt, it wouldn't be long before I was asking myself, *Why is space so hard?*

Physicians on Earth are trained to prevent illness and understand the pathophysiology of disease, how the normal functioning of the body becomes altered by disease. Such an understanding is critical to determine the best approach to managing illness or injury and was a fundamental part of my many clinical responsibilities in the emergency department. The prefix "patho" is derived from the Greek "pathos," meaning "suffering or disease," and reflects a disruption of the normal physiology or functioning of the body in which generally one organ system is affected by disease. Space medicine clinicians quickly understood that the normal functioning of the body is different during spaceflight. The physiological responses or adaptations observed in astronauts are widespread, including changes in cardiovascular conditioning, muscle strength, bone density, orientation and balance. With longer stays aboard the International Space Station, more health effects are being observed and while these changes reflect how quickly the body adapts to the microgravity of space, many are maladaptive from the perspective of living in a gravitational world — whether on Earth or another planet. What happens to the human body in space can adversely affect the health and functioning of astronauts while there, and when they once again experience the effects of gravity.

These adaptations to space, in many ways like those associated with aging on Earth, are reversible when astronauts return home but can also present a challenge to clinicians trying to diagnose disease while they are in space. How does the adaptation to space affect the pathophysiology of disease in that environment? Will symptoms of common illness present themselves differently in space? For instance, in appendicitis the signs of disease are localized to the position of the appendix in the right lower abdomen. Will it be the same in space? For the most part the answers remain unknown, and the challenge for physician astronauts is to work with the flight surgeons in mission control, the experts in aerospace medicine, to diagnose and treat their colleagues.

Physician astronauts have the unique opportunity to understand firsthand the changes that are taking place as their own bodies adapt to space, and on occasion help diagnose space-related illness or maladaptive changes to space in themselves. Such was the case for the first description of spaceflight associated neuro-ocular syndrome by two physician astronaut colleagues, shared in Chapter 2, "Flying Blind." In both of my spaceflights I treated my colleagues for relatively minor medical issues and noted the many changes that took place as I adapted to space and readapted after the mission to being back on Earth.

For me, the changes started the moment after arriving in space, beginning with a mild headache and a puffy face that seemed to increase over the first few hours of the mission, and then remarkably thin legs as my body adapted to the absence of gravity. By today's standard of six-month or longer missions to the space station, both my spaceflights were short — the combined total was around 28 days in space. While I experienced mild cardiovascular deconditioning, there was minimal decrease in my muscle mass and bone density. Incredibly, the absence of gravity most affected my height — in space I was six feet, two and three quarters of an inch! On Earth at my tallest I'm six feet, one inch. The extra inch and three quarters came from the disc spaces widening between the bones in my back and a reduction in the curvature of my spine. Unfortunately, as soon as I stood up on the mid-deck of the space shuttle after we landed, I went back to my regular height. The discs between the vertebrae in my back were squashed back to their normal shape and fortunately they tolerated the return of gravity without failing. Not surprisingly, there have been incidents of herniated discs in astronauts after spaceflight.

The most striking feeling was the vertigo and lightheadedness that I experienced after standing up in the space shuttle after landing back on

Earth. Clutching my spacesuit helmet in its bag, I wondered, "Who snuck a brick into my helmet bag?" then realized that it was all part of being gravitationally challenged — a feeling that resolved relatively quickly over a couple of weeks.

In caring for the elderly, physicians often speak of activities of daily living, those things we do that are a routine part of our day that may become more difficult as we age. To live in space is to live in a world where everything is different. In an environment without gravity, simple things like brushing teeth, sleeping, eating and going to the bathroom require a new approach. NASA does an excellent job training rookie astronauts for all aspects of living in space (including toilet training), but there's nothing quite like being there to discover just how challenging life can be outside the pull of Earth's gravity.

Excitement, elation, exhilaration — there's no word that adequately captures the feeling of waking up on Earth knowing you're going to go to sleep in space that night. After an eight-and-a-half-minute ride going from being stationary to travelling 25 times the speed of sound, the crushing forces of acceleration are replaced by the graceful motion of floating to look out the window at the distant Earth. Learning to move with fingertip forces is critical — no one wants to be the "bull in a china shop" bouncing off the walls of the spacecraft and causing collisions between colleagues. Just as novice sailors must get their sea legs, astronauts must find their space arms as they adapt to a world without ceilings or floors.

In the same way that music touches the soul, the experience of viewing Earth from afar through the windows of the spacecraft can be overwhelming; and going on a spacewalk, where being outside is to be immersed in the majestic beauty of our planet, brings out emotions that few have experienced. The awestruck wonder of that first glimpse of Earth seems like it will last forever. But for most long-duration astronauts, at some point the novelty of the microgravity experience wears off and the effects of living in isolation begin to emerge.

Adaptations to space travel have as much effect on mental and emotional well-being as they do on physical health. President Kennedy understood the many difficulties when in May 1962 he set the goal for the early NASA program to go to the Moon, saying he took on such challenges "because they are hard." Why *is* space so hard? That question prompted us to write this book, to bring to light not only the many challenges but also to revisit the promise of space travel.

Despite the difficulties and dangers of exploring the extreme, harsh environment of space, humans are a spacefaring species. We have lost neither the curiosity nor the wonder for what lies beyond. The spacefaring nations are already planning lunar return missions led by NASA in the hope of one day flying to Mars. Meanwhile the accessibility of space has increased with the emergence of private-sector companies dedicated to taking a new generation of astronauts to live on space stations in low Earth orbit, and potentially to visit the Moon and eventually colonize Mars.

This may be the millennium when humans live on other planetary bodies in our solar system. Understanding how they adapt to the very different new environments of partial gravity and microgravity, keeping them healthy, and learning about the clinical issues associated with living in space are important topics for the future of human space travel. Space medicine is also teaching us valuable lessons about how we can improve human health here on Earth. Some of the strategies for keeping astronauts healthy in microgravity can be applied to seniors and more broadly to enhancing medical care in remote communities. The now widespread dream of human space travel inspires and continues to fuel our curiosity about what it's like to explore this final frontier, and we hope this book will provide insight into some of the challenges of living in space — and even help to improve life here on Earth.

DAVID R. WILLIAMS

CHAPTER 1

The Adaptation Advantage

N ASA astronaut Christina Koch pulled off her space helmet and clasped her hands, smiling in the strong glare of the desert sunshine in Kazakhstan.[1] Beneath her was the Soyuz capsule that brought her back from space, where she had just completed a record-breaking 328 days of service as an astronaut on the International Space Station.

A small crowd of supporters surrounded her Soyuz on February 6, 2020, cheering and waving their hands as Koch flashed a double thumbs-up. You would think that having reached the end of a long journey, she could now take the time to celebrate at some party with friends, but in reality her voyage was nowhere near finished.

A video from NASA Television[2] shows Koch moving carefully on top of Soyuz before three Russian helpers delicately brought her to the ground. Those first few moments back on Earth are usually tough on astronauts, even at the high level of fitness they have achieved through rigorous training and despite Koch's young age of 41. For almost a year, she floated in space like a superhero. Now, Earth's gravity was exerting its force on her limbs and her head. If estimates from past astronauts are true, she would need to spend almost a year in rehabilitation to feel fully like her own self — at least one day of recovery for each day in space.

Six days later, Koch spoke with reporters from NASA's Johnson Space Center in Houston[3] about how the long road to feeling "normal" again was going. "I feel great," she said, calling herself fortunate for avoiding the problems with motion sickness that plague many astronauts upon return.

"But what I have noticed," she continued, "is that my balance has taken a little while to get used to. The physical act of walking was something to get used to. . . . I think just all of the new things that I'm experiencing, you see your mind wake up to the sensory experiences that define Earth and the things that are here."

Koch is not alone in her struggles to readjust to gravity and life back on the Blue Planet. British astronaut Tim Peake once called his sensations on Earth, after returning in 2016 from just 186 days or about six months in space, the "world's worst hangover."[4] Canadian astronaut David Saint-Jacques, who admitted to feeling nauseated after his return from six months aloft in 2019, joked that "gravity is not my friend."[5]

The first few minutes on Earth are notoriously tough on astronauts. For research purposes, NASA and most of its international counterparts ask astronauts to try walking (in a private tent, away from the television cameras). A doctor is always right beside them, because this first sortie on foot is a slightly dangerous one. The astronaut could faint "due to [a] mixture of reduced ability to control their blood pressure, heart changes and dehydration," reported one European Space Agency account from a medical doctor responsible for astronaut health.[6] They feel seasick and have difficulty orienting themselves, especially if they rotate. Bones and muscles are especially prone to injury, including those in the back, because there's little standing going on in space. In space the postural muscles atrophy, becoming weaker without the weight-bearing and resistance demands of an environment with greater gravity. The discs between the vertebrae — the bones that make up the spinal column — are particularly vulnerable when astronauts stand up for the first time after a mission. Herniated discs can happen to any of us when lifting a heavy object the wrong way, because of the abrupt pressure from bearing additional weight; similarly, the sudden increased load to the back when they first stand up has caused disc herniation in some astronauts.

Learning how to help astronauts readapt to gravity is important when you consider the challenges of going to other worlds in the solar system. On Mars, there may be no helpers available to assist astronauts out of their spacecraft when they land on a planet that, although it has only 40 percent the gravity of Earth, will challenge them as they struggle to stand up after a voyage of several months in microgravity to get there. Preparing astronauts to be safe, injury-free and active when they return to Earth has important implications for getting future astronauts ready for all of the things they will have to do to set up bases and establish a productive work environment after landing on the Red Planet.

Space affects astronauts in ways you would not expect, as they may be more prone to infection and illness because immune cells behave differently in space. Research in this area is still emerging, but it means that — even if there is no pandemic — astronauts must observe a strict quarantine before and after flight so as not to chance illness.

NASA and its international partners hope to reach the Moon again in the late 2020s, and perhaps voyage on to Mars later this century. But going on such long journeys seems chancy to some, given what we know of the challenges astronauts face when they come back to Earth after missions lasting only a few months on the International Space Station (ISS): feeling light-headed when they stand up, weak and unsteady when they walk and disoriented when they move their heads. One of the defining questions arising from the most recent space station missions is how researchers and doctors can develop effective countermeasures so that healthy astronauts in their thirties, forties and fifties can quickly readapt to living and working in a world with gravity. What will help astronauts, cosmonauts (from Russia) and taikonauts (from China) take on future long-duration missions, including long voyages to the Moon, Mars and beyond, with the least medical risk as they adapt to the absence of gravity en route and partial gravity at their destination?

"Countermeasures" are the physical exercises astronauts do, and the medications or devices they may use, to adapt to spaceflight and prepare for returning to Earth. Fortunately, more than 20 years of research has allowed doctors to begin developing an adaptation advantage, creating programs tailored to each spaceflyer, to keep them in shape in space and to ease their arrival back on Earth. We may not have all the answers yet, but doctors are confident that modern medicine can help manage many of the challenges Koch and her fellow spaceflyers face after long stints in space.

Countermeasures developed for space travel also have potential application here on Earth. Doctors have noticed that many of the changes seen in returning astronauts — deconditioning, bone loss, balance and gait disorders — are like those associated with advanced aging, yet for the astronauts they are reversible. Are there lessons from space research that might help all of us as we get older? There are some initial thoughts: keeping active as we age is important to reduce bone loss and muscle weakening, in-bed exercise programs are better than no exercise and swimming is a rehabilitation technique that works well for astronauts and helps those with joint pain maintain their fitness. Perhaps one day there will be other countermeasures used by astronauts that help mitigate the effects of the aging process.

When the body goes to space, just as in any other environment that differs from the norm, it is trying to adapt itself. If our body couldn't adapt to new conditions, we wouldn't survive in some of the less hospitable environments on Earth. Think about how your body sweats when you enter a hot sauna, or how blood flow to your fingers and ears reduces if you step out of your house into cold winter weather. These are all normal and reasonable adaptations to survive on planet Earth — and in most cases, our body adapts automatically.

When an astronaut leaves Earth, their body rapidly changes in response to living in a radically new environment. Most astronauts spend two to two and half years in mission-specific training before taking on a new six-month (or longer) stay aboard the space station, not to mention the years they may have spent supporting other missions. They understand the basics of how the body changes and veteran astronauts understand how to adapt to space; but what is unpredictable to doctors and scientists is how each individual adapts and how that adaptation changes when astronauts fly missions at different ages over the course of their career.

Adaptations to space start happening within hours or days. Many astronauts report a "stuffy head sensation" and in some cases "slight motion sickness" within the first couple of days of spaceflight, explained Andrea Hanson, a NASA medical doctor and the ISS exercise countermeasures operations lead, in a 2018 interview:[7] "The body responds pretty immediately to living in the microgravity environment . . . going through a series of adaptations to adapt to this new sensation of free float." The rapid transition to microgravity causes fluid to shift from the lower extremities to the face, torso and upper extremities. This causes a stuffy nose like that experienced with a cold, alters the taste of food and creates a somewhat humorous "moon-faced" appearance that can be more striking in some astronauts than others. If nothing else, it's a great way to get rid of facial wrinkles!

Beyond the frequent cold-like symptoms, Hanson added, the spine begins to elongate since there is no gravity compressing it, making astronauts taller than they are on Earth. The body also realizes that the sensations of "up" or "down" no longer matter in space, changing how astronauts orient themselves in this new environment. "All those little mechanosensors that tell your body it needs to keep strong to even stand up straight, go up and down stairs, sit in a chair, and stand up again, they start to turn off," Hanson said. "That's where we start to see the muscle strength losses and bone atrophy that we know our astronauts can experience, even at as soon as two weeks in space."

One to two weeks was the length of a typical space shuttle mission, so many of the more serious effects of space travel were not felt on these short trips, but ISS missions are generally between six and 12 months to prepare for future long-duration missions beyond Earth orbit. A year sounds like a long time in space until we remember that getting to Mars and back will likely be a two-year round trip, unless there is a huge change in propulsion technology that lets us travel large distances more quickly. The impact on the human body will be even greater as the astronauts spend longer periods away from Earth's gravity on these extended voyages into uncharted territory.

Science fiction and a few experiments in space show the potential of generating artificial gravity, but such a process would likely be expensive and require a very large, rotating spacecraft that presents significant engineering challenges. So for now, we're stuck with a microgravity environment, and that requires another approach to ensuring astronaut health — the development and use of individualized countermeasures.

Exercise is a very important component in maintaining astronauts' health and strength, and the ISS has countermeasure technology built right into it. For example, NASA has a treadmill on the space station named after comedian Stephen Colbert, who long ago jokingly tried to get his name attached to what we now call the U.S. Tranquility module.[8] NASA appreciated the publicity his show brought to the space program and, after reportedly considering naming the toilet in his honor, compromised by agreeing to name the treadmill COLBERT, a combination of the talk-show host's name with a newer NASA acronym: the Combined Operational Load Bearing External Resistance Treadmill.

COLBERT is a little fancier than your usual gym equipment. The most obvious change is it has special straps to keep you running in place, lest you float away. It also includes data collection devices to evaluate how well the treadmill minimizes bone and muscle density loss, including metrics such as session duration, body loading and treadmill speed. Its maximum speed is 12.4 miles per hour (mph), a deliberately high maximum that's not as fast as Olympic athletes sprinting in a 100-meter race but a lot faster than you would run longer distances. For most astronauts, the range gives plenty of allowance for individual variance, with a typical spaceflyer expected to pace themselves at 4 to 8 mph.[9]

Similar to the COLBERT, NASA has a special exercise bike on board — a Cycle Ergometer with Vibration Isolation Stabilization (CEVIS), which is connected to the space station using wire tethers. Both the COLBERT and the CEVIS have vibration isolation systems that prevent exercise from shaking the space station and its sensitive experiments. Astronauts can use CEVIS to keep their heart rate at a specific target and can change the force with which they push the pedals to vary resistance and maximize the benefit of their time on the bike.[10] If they stay on the CEVIS for 90 minutes, they can join the unique club of those who have cycled around the world in that amount of time, as the ISS orbits the Earth every hour and a half!

There's even a solution for weight lifting in an environment where it normally wouldn't be possible. A piston-driven machine known as the Advanced Resistive Exercise Device (ARED) uses vacuum cylinders and a cantilever device (similar to a teeter-totter) to allow astronauts to lift the equivalent of a load up to 600 pounds at once.[11]

"Essentially, when you're exercising on ARED, you're kind of in a clam-shell or in between a clothespin where you're pushing off against a platform and up against an exercise bar," Hanson explained. "So whether you're standing and you have that exercise bar on your shoulders, or you've installed the bench attachment and you can do bench presses, ARED allows one to conduct over 45 different types of exercises."

ARED is not the first such device to be used on the ISS, as NASA flew the Interim Resistive Exercise Device (iRED) in the early years of space station operations. Although at that point few astronauts were taking on space missions as long as six months, NASA documentation[12] nonetheless noted that the iRED likely did not produce adequate stimulus for muscle strength, muscle mass and bone mineral density compared with the free weights that astronauts use in gyms on Earth.

A 16-week training study conducted by the Exercise Physiology and Countermeasures Laboratory at NASA's Johnson Space Center examined the effects of free weights vs. iRED in a set of healthy earthbound men and women who were described as "untrained" in such exercise. While this study group is not fully representative of the astronaut population — which tends to be highly fit and experienced with exercise devices — the study showed that free weights produced a greater increase in muscle strength compared with iRED. The device also did not change bone mineral density as effectively as free weights. The researchers hypothesized that free weights produce more eccentric loads than the iRED — loads that allow the muscle to work against resistance as it lengthens — which may have accounted for the change in efficacy: "Despite performing resistive exercise with the iRED, International Space Station (ISS) astronauts still experience losses in muscle strength and endurance and [bone mineral density] following long-duration space missions (>4 months)," NASA added. "The peak iRED load level may be inadequate to ease musculoskeletal losses in microgravity where the subject is not required to lift their body weight."

Happily, improvements were notable after the ARED was installed on the space station in 2008. Early studies and an updated one in 2014 showed that as long as astronauts had "adequate energy intake" and took vitamin D supplements, use of the ARED resulted in bone mineral densities "virtually unchanged from pre-flight" — a huge win. Bone resorption was still occurring, but bone formation increased with use of ARED.

"These data are encouraging and represent the first in-flight evidence in the history of human space flight that diet and exercise can maintain bone mineral density on long-duration missions," NASA wrote.

Although the quantitative metrics were satisfactory, the qualitative results may have been less positive. It's possible that despite maintaining bone density equivalent to acceptable levels on Earth, the quality and strength of that bone may not be the same. Would the astronauts still be more susceptible to fractures and future bone loss? Even with countermeasures in space, we cannot always be sure of the long-term effects of this unnatural environment on the human body.

It is useful to note that the perennial problems plaguing scientists when it comes to studying astronaut populations are the small sample size and lack of diversity in the subjects. After 60 years of space travel, fewer than 600 people have flown into orbit as of this writing; of those, only a limited number have spent as long as six to 12 months aboard the space station. Most people who fly to space are in their thirties, forties or fifties and, for now, the population trends to white males, although newer astronaut classes include more women and more ethnicities, which will help balance the subject population.

Regardless of their gender or background, all astronauts are in excellent physical and mental health before flight and are motivated to exercise and to remain fit — which may not necessarily reflect the typical fitness, attitude or practice of a suburban human stuck in an office chair for eight hours a day. However, the lessons learned from developing an adaptation advantage for astronauts can help to improve the health of the average person here on Earth. Long-duration spaceflight has confirmed the importance of fitness in building resilience and the role of individualized wellness plans. And the idea of preconditioning astronauts before a long-duration mission — through aerobic and resistive exercise to optimize cardiovascular fitness, muscle strength and bone density before they get to space — has direct applications for individuals undergoing elective surgery, who might benefit from "prehab" to keep hospital stays shorter and reduce the possibility of post-operative complications.

As we saw earlier, astronauts in orbit for long periods of time experience outcomes similar to seniors on Earth, especially those who are chairbound or bedbound. (NASA, in fact, uses bed-rest studies from volunteers to make comparisons to astronaut health.) For instance, studies in space have shown that astronauts' arteries stiffen, aging the vessels by an equivalent of 20 years on Earth for every six-month or so period in space. Researchers and space medicine physicians hope to apply the lessons learned from human spaceflight to help understand and potentially manage some of the physiological changes associated with aging.

In Canada, for example, the Schlegel-UW Research Institute for Aging in Waterloo, Ontario, is involved in a space study called Vascular Echo and is directly applying the results to a seniors' residence on Earth where the institute conducts regular studies. Principal investigator Richard Hughson and his international team are collecting cardiovascular data from space station astronauts that may benefit geriatric patients in an attempt to slow down or reverse the stiffening of the arteries associated with aging.[13]

We know that exercise is key to maintaining human health and to preventing all kinds of chronic diseases, including type 2 diabetes, heart disease, cancer and Alzheimer's. Space studies like this one reinforce just how critical it is to astronauts and Earth-dwellers alike. Exercise is so important to the adaptation advantage that astronauts in space are scheduled for 2.5 hours of exercise time a day for medical countermeasures and mental health. That doesn't mean they're hitting the bike, the treadmill and ARED for all that time, as there is some setup required for the various devices. But a typical astronaut will log 90 to 120 minutes six days a week. A more motivated one might choose to, say, run a marathon in space for an extra challenge, like Sunita Williams[14] or Tim Peake.[15]

What is important to doctors is not so much the few astronauts who run marathons, but the promise from NASA scheduling managers that no astronauts will be permitted to delete exercise time from the manifest — unless, of course, a true emergency intervenes. In a multi-month mission, this exercise time is crucial to keeping them healthy, and to helping maintain their bone and muscle strength until their return to Earth.

Given that a typical full space station crew includes up to seven people, scheduling daily workout time for each of them can be a challenge. Russian cosmonauts have some exercise equipment of their own, but all the same, space station schedulers need to take into account the exercise needs of six or so people, planned equipment maintenance and downtime, and all the other activities that crew members must do — including experiments, taking care of space station systems, food preparation and rest time.

Exercise plans are also highly personalized to individual astronauts. Each one is typically assigned astronaut strength, conditioning and rehabilitation specialists (ASCRs) who look at the protocols for exercise before flight, during flight and after flight to make sure that the workouts are optimized for each person's body.

"They [ASCRs] check in on a regular basis to make sure that crew are comfortable with the exercise hardware, if they have any concerns," said

Hanson, the NASA countermeasures specialist. "Maybe they know that they have a really packed schedule coming up for a couple of days, and they need to make some trades in their exercise workout to make sure that they have the full mental aptitude and are prepared to take on the otherwise stressful schedule and balance that with use of exercise."[16]

Data on these various exercise devices continues to be collected as NASA targets long-duration missions elsewhere in the solar system that will require new or adjusted countermeasures for successful human adaptation. Early in Joe Biden's presidency, his administration committed to continuing the Donald Trump–era Artemis program, which had been shooting for human Moon landings in 2024. Although Biden has not committed to a timeline by the time of this writing, NASA is working on numerous plans to keep its astronauts healthy in microgravity and will work to make even more effective devices for these lunar missions and for eventual landings on Mars. "That's what we're really trying to do — taking all those lessons learned from that hardware design, and the stresses that have been put on that hardware, to make sure that the devices being designed for those smaller exploration vehicles are going to be able to stand up to the stresses and the continuous use that we expect that they will have," Hanson said.[17]

Hanson hinted that in tight quarters of lunar spacecraft or bases, it may be that a single device could replace the three main ones that astronauts use on the space station. "We're being challenged to . . . have one piece of hardware that's going to be a cross-training, both aerobic and strength-training, device. We know that multimodal exercise is important, which is why we have the cycle, the treadmill, and the strength-training device today." Thankfully, in the case of lunar missions the partial gravity will provide a degree of conditioning.

Medical countermeasures in orbit are not just limited to exercise. Keeping crew members mentally healthy for six months or more at a time is essential. In space, long-duration astronauts have had to deal with the emotional fallout of major events, ranging from national tragedies such as the terrorist attacks of September 11, 2001, to the deaths of family members.

Astronaut Scott Kelly was in the midst of a six-month mission when his sister-in-law, then-Congresswoman Gabrielle Giffords, was shot and severely injured during a community event in Tucson, Arizona. From space, Kelly described how he worked to support his twin brother, Mark (also an astronaut), during this time: "The best I can do is continue to do my job as I've done all along, support my brother and my kids the best way I can by means onboard the space station," Kelly said during a January 2011 news conference[18] from space, adding that another way he was working to support his family was by leaning on his crew. "The way we deal with challenges in this environment is through teamwork," he said.

In most cases, astronauts get along just fine with fellow crew members. That said, interpersonal difficulties or frustrations from feeling overloaded — in the rare cases they arise — must be dealt with on the spot so the issue does not compound across the mission. To their advantage, crews spend so long training together in difficult circumstances on Earth, such as remote locations or underwater, that they are accustomed to being in a small group in close quarters, under time pressure and unusual stresses. Training builds teamwork that pays so many dividends even before going to orbit.

Most of NASA's astronaut corps has at least some experience in isolated environments. Koch, for example, used to be a telescope researcher in Antarctica — a relatively common location for astronaut candidates to have visited during their careers before applying to NASA. In a 2019 NASA interview,[19] Koch said Antarctica is an excellent space analog: "I would say [it's due to] the harshness of the environment, and the mental and physical kind of fortitude that it takes to be successful somewhere like that, coupled with the science that you can do in that unique environment."

Koch added that her role in Antarctica as research associate mimicked what astronauts often do in space. While the public may imagine these explorers as people who strike out on their own most of the time, in reality, Antarctic and astronaut researchers alike work in tightly scheduled environments in close quarters and in constant communication with teams hailing from other communities and countries.

To consistently monitor their mental health in space, all astronauts on the U.S. side of the space station have regular sessions with a psychologist to discuss any concerns that may arise. This psychologist can work with the mission control team to help the astronaut adjust their workload, request time off or ask for more assistance from ground teams. Protocol also gives all orbiting astronauts a dedicated personal support team, including other astronauts on the ground who represent them in meetings at mission control and regularly check in on any close family members such as spouses or children.

During her first NASA press conference after landing, Koch discussed personal countermeasures[20] she had taken to help with her own mental health during nearly a year of spaceflight, describing her approach as "mental cheerleading," including "put[ting] something on repeat in your head that's going to be constructive."

"For me," Koch added, "that was always focusing on what I had, and not the things I didn't have; focusing on the unique aspects of my life that one day, I would just wish I could have back for a second. And every time you think about it, from that perspective, you're done feeling like you miss home. You're just ready to take on the day."

The medical team working with each astronaut can also prescribe drugs or adjust prescriptions as the situation arises. For example, the first few days in orbit often see astronauts pop anti-nausea pills, even if only as a precaution, since there is so much to do after arriving at the International Space Station. Another common short-term prescription for astronauts is sleep medication, since the stresses of astronaut schedules and the relatively noisy and bright conditions of a space station can interfere with good sleep. Again, astronauts have a lot of training in sleep deprivation even before flight, not least because they must travel all over the world for their training and deal with time zone changes.

Still, floating in space and attempting to sleep in an unfamiliar environment is a difficult mental shift, at least at first. A 2018 study from the journal *Military Medical Research* examined numerous sleep logs and drug allocations among space shuttle and space station missions and concluded that many astronauts (although not all) have used drugs to assist with recovering from sleep deprivation,[21] at least in the short term. Other countermeasures have been made as well; for example, in August 2016, NASA had astronauts install new lighting on the space station to replace its old fluorescent light bulbs with adjustable light-emitting diodes (LEDs), allowing crew members to change the intensity and color of the lighting as required.[22]

After landing, ASCRs provide guidance to make sure that each astronaut regains their balance and strength, so that within a few weeks they can take on light work duties. Within a few months they are back to typical activities like driving and running. ASCRs must also keep in mind metrics such as age, how much exercise the astronaut is taking on, and how to help the astronaut keep exercising during a busy postflight schedule that includes debriefs and public relations.

Typical crew members, Hanson said, can walk on their own a day or two after landing. Full muscle strength tends to come back within two to six months after a six-month mission. More concerning are other physical issues that are not as visible; for example, bone density tends to take longer to recover than one's sense of balance.

Experiments on an astronaut's health continue well into the recovery period, to compare how they are doing after spaceflight with during or before. Many astronauts voluntarily come back to NASA each year, even after retirement, to take part in long-duration studies on astronaut health.

One of the more famous reports is the "Twins Study" on the Kelly brothers, when Scott returned to space for a year during his final mission and Mark, then retired, took part from Earth. A combined summary paper of the research was published in the journal *Science* in 2019, representing the work of hundreds of scientists around the world who had been tracking the twin astronauts even before Scott Kelly flew to space for the last time in 2016–17. The study was wide-ranging: its investigation teams were looking at matters from immune system response (Scott successfully received the first flu vaccine in space) to astronaut microbiomes, gene expression and the developing science of fluid shifts.[23]

NASA said it will continue to develop countermeasures for astronauts, taking into account the results of this study, but it also pointed to how far the science of taking care of people in orbit has come in the 20 years or so of space station research: "NASA has a rigorous training process to prepare astronauts for their missions, a thoroughly planned lifestyle and work regimen while in space, and an excellent rehabilitation and reconditioning program for them when they return to Earth," the agency said. "Thanks to these measures and the astronauts who tenaciously accomplish them, the human body remains robust and resilient even after spending a year in space."[24]

Getting ready for space and recovering from its effects will always be a tough business. Psychologically there is the challenge of working in a small, confined environment far from the comforts of home and family; issues such as crew dynamics and workload must continue to be explored and trained for to keep spaceflyers happy under these tougher-than-usual conditions. Physical matters are also of concern, especially declines in bone mineral density and muscle strength, as well as other effects we'll examine in later chapters. With robust countermeasures in place, we are seeing encouraging progress on all fronts. And ongoing research, like the Twins Study, will allow NASA to continue to monitor its astronaut population for potential long-term effects and conditions, and to develop exercises, psychologist interventions, drugs and other countermeasures to address emerging issues.

Even after 20 years of research on long-duration missions on the International Space Station, and nearly 60 years of human spaceflight in environments ranging from low Earth orbit to the Moon, there are still enormous challenges in space medicine, and we still have much to learn. Despite the ability of humans to adapt to gruelling conditions in space, and the advantage space travellers gain from breakthrough countermeasures that help them cope in microgravity and back on Earth, space is indeed the final frontier — pushing the body and the mind to extremes and constantly challenging whatever adaptation advantage we humans have. Future chapters will dive into various health issues in more detail, revealing some of the things we've learned about ourselves thanks to space travel and space research, and show some of the things we're still trying to grapple with before expanding further into the universe.

CHAPTER 2

Flying Blind

The Expedition 20/21 mission to the International Space Station had been proceeding well, the crew of six busy with the multidisciplinary science experiments, robotic operations and routine maintenance of the complex systems that are required to keep the world's only orbiting laboratory functioning properly. But as the mission progressed, NASA astronaut Mike Barratt and Canadian astronaut Bob Thirsk began having difficulty reading the electronic procedures for the tasks they had to perform.[1] As physicians, they were both curious to understand the cause for this change in their vision. They were suffering from what is now known as spaceflight associated neuro-ocular syndrome (SANS).[2] Researchers suggested it could be caused by the shift of body fluids that takes place living and working in space.

Roughly 60 percent of the human body is made up of water — it's in our blood, tissues and organs. On Earth, the distribution of that water is governed by the force of gravity. Lift an arm above your head and you can feel the blood draining and see the veins collapsing as the blood returns to the heart, following the gravitational gradient. One of the earliest changes that affect the body in space is something called "puffy face–bird leg" syndrome: without gravity, in space there is a shift of about a liter of fluid from each leg back to the upper body. This redistribution of fluid causes the face to appear puffy and the legs to look very thin. Those fluid shifts suggested to researchers that the quest to understand SANS should begin with a study of the circulatory system.

From the time we first start walking, we learn about gravity every time we fall. We become accustomed to picking ourselves up, dusting ourselves off and trying again. By the time we make it to elementary school we can relate to artist Gerry Mooney's quote, "Gravity. It isn't just a good idea. It's the law."[3] Soon the rules associated with classroom conduct become second nature and show that gravity affects blood flow in the body: if you have a question you raise your hand, and there's no sensation quite like the feeling of numbness that accompanies keeping your hand above your head as you wait to be seen by the teacher — especially if you need to go to the bathroom. There's then a sense of relief as you lower your arm and feel the reassuring flow of blood back into the limb. We also notice it when we stand up too quickly and feel dizzy — or wonder why our shoes seem so tight when we try to put back them on after a long airplane flight.

The gravitational impact of staying in a prolonged inflight seated position results in blood pooling in the legs. Leg exercises can help reduce this effect, as the contracting muscles help push blood from the legs back to the heart. Getting up and walking around also helps, and the lucky business-class travellers can prevent the problem by reclining their seats and bringing their legs up in front of them. But it is not just the position of the legs; the lower pressure of the aircraft cabin can also cause or increase the amount of swelling. Compression stockings help but it can still be a challenge getting those shoes back on after the flight.

If gravity can have so many effects on the distribution of fluids in our bodies, what happens to other animals whose cardiovascular system is influenced by the force of gravity? What about the giraffe, whose 6-foot neck helps it graze on leaves high in the trees? Why doesn't it get dizzy or have trouble seeing after drinking at the water hole, then moving its head vertically by 15 to 20 feet or more?

Evolutionary biologist David Barash has studied how giraffes have adapted to their particular niche and notes, "The name 'giraffe' comes from the Arabic word 'zarafah,' which means 'one who walks swiftly,' and giraffes use those long legs to walk swiftly indeed. . . ."[4] In addition to speed, they have also developed a few adaptations to help them overcome the issues associated with their height — it would be a problem if they fainted every time they lifted their head to the surrounding trees to eat.

"To pump blood seven feet above the animal's heart to that towering head, an exceptionally high blood pressure is required, as much as three times the human systolic level," Barash said. Additionally, "to prevent blood from pooling in their feet, which are at the end of some very long legs, giraffes have evolved the equivalent of compression stockings, like those that people use post-surgery or to prevent deep vein thrombosis on long airplane rides. The giraffe's hack consists of highly elastic blood vessel walls, combined with an extensive capillary bed. By restricting perfusion of fluid into surrounding tissues, these structures keep a giraffe's blood in its vessels, where it belongs, rather than in surrounding tissue."

Essentially, the giraffe has a built-in G-suit. Extremity compression has been used by pilots and astronauts for decades to help them avoid fainting when G-forces, which amplify the effects of gravity, cause blood to pool in their legs. The first G-suit was invented by University of Toronto professor Wilbur R. Franks in the early years of World War II,[5] and derivations have been in continuous use since. Space shuttle astronauts routinely use a G-suit during re-entry and after landing to stop them from fainting. The antigravity suit clearly helps prevent fluid pooling in the legs, but what about fluid shifts to the head? Could understanding what happens when a giraffe drinks provide clues to the cause of SANS?

When giraffes visit the watering hole, they spread their front legs far apart, lowering their body and reducing the distance their head will drop. This decreases the gravitational gradient that causes blood to pool in the neck veins. They also have valves in the veins of their neck to stop the reverse flow of venous blood back to the head, and a "specially adapted (tissue) compression system in their necks, which prevents too much blood from rushing to their heads when they bend down to drink."[6]

Humans do not share the same adaptive mechanisms used by giraffes. Generally, this is not a problem as humans typically spend more time standing on their feet than they do standing on their head. But there are those who have explored the potential benefits of an inverted body position. Salamba sirsasana is a yoga pose whose practitioners suggest calms the brain, relieving stress and mild depression.[7] More commonly known as a headstand, it is thought to help by increasing blood flow to the brain.

The human brain weighs around three pounds and uses 20 percent of the oxygen carried in our bloodstream. Approximately 75 percent of the brain is water.[8] It sits inside the skull for protection and is bathed in cerebrospinal fluid. If most of the brain is water, perhaps it might be vulnerable to fluid redistribution associated with different body positions. The production of cerebrospinal fluid is closely monitored and regulated by a control system that prevents pressure from building up around the brain. There are also other control mechanisms that regulate blood flow to the head and brain in different body postures. Surprisingly, those mechanisms have been shown to reduce blood flow to the brain during a headstand — so the reported benefits of sirsasana apparently are not based on increased cerebral blood flow.[9]

Using a technique known as pulsed wave Doppler ultrasound, researchers found the diameter of the artery supplying the brain remained "almost unchanged" in this yoga posture. Pulsed wave Doppler is based on the principle that moving objects change the characteristics of sound waves, something we've all experienced listening to the sound of a siren as an ambulance or police car rushes past. By sending short rapid pulses of sound to a blood vessel, it is possible to measure the velocity of blood and follow the changes in blood flow as they occur.[10] Despite the general assumption that a headstand would increase blood flow to the brain, the Doppler tests showed that it actually decreased. The reduced blood flow during the headstand returned to normal after standing.[11]

Experienced yogis and yoginis may not agree with the Doppler test results because of their experience of feeling blood going toward their head when they are upside down. Most of us have experienced the sense of facial fullness, distended neck veins and the red face associated with being upside down, and some even find it uncomfortable. But the Doppler findings apparently disagree with our own experience. The explanation lies in the difference between arteries and veins.

The arteries in our body carry blood away from the heart to take oxygen to the tissues and organs. They have muscular walls that help with blood flow and are an important part of maintaining our blood pressure. In contrast, veins are thin-walled structures that carry the blood back to the heart, where it is pumped to the lungs to be reoxygenated. The cycle continues as the oxygen-rich blood then returns to the heart and is pumped back into the arteries to deliver essential oxygen to the brain, muscles, organs and tissues that make up our body.

Unlike the giraffe that has special valves in the neck veins to prevent the backflow of blood toward the head when it is drinking, blood tends to pool in the neck veins when humans are inverted. If blood pooling in our legs during an airplane flight can cause our feet to swell, what happens to our head when we are upside down? Could doctors find clues from sirsasana practitioners that might help us understand the visual changes in astronauts?

A group of researchers in India published a report documenting changes in the eyes of 75 volunteers in the sirsasana position. Measurements of the eye and the pressure inside the eye were recorded before, during and after the inverted pose, and the results were striking. There was a twofold increase in the pressure inside the eyes of the volunteer test subjects who maintained the posture, in all age groups.[12] The findings were attributed to the redistribution of fluids to the head while the subjects were in the inverted pose.

It is difficult for experts in space medicine to study the many changes that take place in astronauts' bodies when they are flying in space, so over the years they have used different approaches to studying what might happen in space by using a variety of test conditions on Earth. One of these is head-down tilt bed rest, which produces a headward shift in fluids like that found in astronauts in space. Prolonged head-down bed rest has been shown to produce changes in the human eye as well as in bone density and muscle strength.

Researchers in the NASA Flight Analogs Research Unit, located at the University of Texas Medical Branch in Galveston, were interested in assessing whether head-down tilt bed rest and other detailed tests might be useful in understanding not only the changes in eye pressure, but also why astronauts were developing changes in their vision in space. They performed detailed eye examinations on 16 subjects and found a small myopic shift, an increase in intraocular eye pressure and thickening of the retinal nerve fiber layer at the back of the eye.[13] Despite the changes, the authors concluded that the results "did not produce clinically relevant visual changes" and suggested that further research is required to determine if prolonged head-down tilt bed rest can produce eye changes similar to those found in astronauts.

With SANS listed as one of the critical issues to be resolved for long-duration missions beyond Earth orbit, the studies in astronauts have continued and further work has been done evaluating the usefulness of the head-down tilt model. Astronauts are routinely evaluated before, during and after spaceflights with a technique called optical coherence tomography, which allows clinicians to assess edema (swelling due to fluid accumulation) and changes in the thickness of a layer of nerves called the choroid that forms the middle layer of the wall of the eye.

Using a combination of sustained head-down tilt (previous studies had allowed the subjects to sleep with their head propped up on a pillow) and slightly elevated levels of carbon dioxide similar to those found in the air on the International Space Station, researchers were able to demonstrate changes after 30 days of head-down tilt that were similar to those found in affected astronauts.[14] The study is of particular interest as it is the first model of reversible optic disc edema to study SANS on Earth that might assist in developing countermeasures to help prevent the problem in space.

One of the countermeasures proposed by the researchers was to use a technique developed in the Russian space program called lower body negative pressure (LBNP). This requires the user to put on a specially designed suit that creates a negative pressure around the legs, literally "sucking" fluids back down to the lower extremities (sort of the reverse of a G-suit). In space it would emulate the effects of prolonged standing on Earth and might relieve the facial fullness and thin legs characteristic of puffy face–bird leg syndrome. While there are suggestions that LBNP may be of benefit to counteract fluid shifts, it is unclear how long the benefit lasts when it is stopped. Given that astronauts must be tethered to a particular location on the space station when the device is in use, and using it requires continuous cardiovascular monitoring, it doesn't seem to be a practical solution for sustained use.

Despite the limitations of LBNP, it has been assessed as a SANS counter-measure. Although it has been shown to significantly lower the increased intraocular pressure associated with head-down tilt bed rest, it did not decrease the thickening of the critical nerve layer at the back of the eye. Perhaps the engorgement of this tissue during spaceflight is due to chronically reduced venous drainage within astronauts' heads during long-duration missions.[15] Fortunately we don't spend a prolonged amount of time upside down on Earth, and the discovery of SANS is an important reminder that despite the adaptability of the human body to living in space, we are more at home when under the influence of gravity.

Most of us recall looking up into the night sky as a child, awestruck by the seemingly endless number of stars. Our innate curiosity causing us to consider what's up there, what it would be like to explore the vastness of space. Wondering are we alone or is there life on some other planet in our galaxy or elsewhere in the universe? The unparalleled capability of modern telescopes has given us the ability to see detailed images of distant stars and planets, providing a unique perspective on the vast final frontier of space. Perhaps it is appropriate that astronauts hoping to one day leave Earth orbit and travel back to the Moon and beyond are becoming somewhat farsighted in space. While SANS may be a challenge for day-to-day life, it is fortunate that their distance vision is preserved as they gaze at the Moon from their perch on the International Space Station, marvelling at the beauty of Earth and sights few humans ever get to see while thinking about what the future of space exploration might look like.

CHAPTER 3

Striking a Balance

"**W**e have serious problems here," Gemini 8 pilot Dave Scott said to commander Neil Armstrong. "We're tumbling end over end." Forty-five minutes earlier Armstrong had performed the first ever docking of two spacecraft, attaching the Gemini capsule to the Agena target module and letting mission control know, "Just for your information, the Agena was very stable and at the present time we are having no noticeable oscillations at all."[1] But shortly after docking, Scott noticed they were slowly starting to move.

"Neil, we're in a bank," he calmly announced. Armstrong was able to use Gemini's thrusters to stop the tumbling. However, the roll immediately began again. With Gemini 8 now out of range of ground communications, the crew had to find a solution, and quickly, as the rate of rotation of the two spacecraft was rapidly increasing. They calmly worked together to find a solution despite the exponentially growing risk of losing consciousness from the forces associated with the uncontrolled motion. During the post-landing press conference, Armstrong described the situation: "Looking at our attitude indicator, we found that our attitudes were indeed changing in yaw and roll. We attempted to bring the combination under control and reduce the rates. Then the problem again uncovered itself, and the rates began to increase to the point where we felt the structural integrity of the combination . . . points was in jeopardy."[2]

With a catastrophic failure of the docking attachment between the two spacecraft an imminent threat, Armstrong was worried about the risk of collision once they performed an unplanned detachment. "We'd reduced the rates to a point where we felt the undocking could safely be performed without a chance of recontact between the vehicles. Since there was rotation in essentially all directions, we wanted to be assured that we could get far enough, around, away from the Agena before some recontact between vehicles was encountered."[3] With precision timing, Armstrong and Scott undocked from the Agena module. But the problem didn't stop.

While Armstrong tried to bring Gemini 8 under control, Scott reported to mission control, "We're disengaged from the Agena. We have a violent left roll here at the present time and we can't turn the RCS's (reaction control system) off, and we can't fire it, and we certainly have a roll. . . ." Without the added mass of the Agena, the roll rate of the Gemini spacecraft rapidly increased, getting close to one revolution per second. Armstrong reacted immediately by turning the thrusters off and using the re-entry control system to stop the spacecraft rotation. It was a disappointing end to an otherwise successful rendezvous and docking, but it could have been a lot worse. A collision could have resulted in the loss of the crew and vehicle, essentially eliminating the possibility of achieving the goal of the future Apollo program: landing on the Moon. By the time Armstrong solved the problem, both astronauts were experiencing vertigo and the physiological effects of the tumbling acceleration.[4]

After the ill-fated 1966 mission NASA corrected the stuck thruster problem on the Gemini craft by changing the ignition system wiring. However, the ever-present risk of vertigo and disorientation in space proved a more complex problem to solve.

The data from Gemini 8's onboard flight recorders showed that the astronauts had experienced a tumbling of around 50 rpm that generated rotational forces of 0.92 Gs for 42 seconds before the spacecraft was brought under control. Were it not for the quick response of the crew, the tumbling would have lasted longer and would likely have incapacitated both astronauts, possibly with deadly consequences.[5]

Most of us have experienced the disorienting dizziness associated with spinning in a circle at some point in our childhood. Adults typically dislike the feeling of twirling and tumbling, but feeling the world spinning around can be fun for kids. Whether it is being swung around by the arms or simply spinning in circles on their own and tumbling to the ground, children seem to love the giddy high. In fact, researchers[6] now feel that children need the dizzying input that comes from spinning, rolling and swinging to assist in the development of different sensory systems and motor skills. These movements may help the child's nervous system create the elaborate connections needed for more complex tasks. But dizziness in space could be disastrous.

Pilots and astronauts must learn when and when not to rely on their sense of balance, hearing, visual cues, and feedback from pressure on different parts of their body to safely operate complex aircraft and spacecraft. There are times when they can rely on sensory input and other moments when those inputs are not correctly portraying what is actually happening. Unquestionably, the Gemini 8 crew had to rely on their training and their interpretation of the data from the spacecraft's systems as they struggled to successfully control the capsule while suffering the effects of vertigo.

Vertigo is an abnormal sense of motion often described as a sense of spinning or loss of balance that can become incapacitating. It can be difficult to appreciate the impact of vertigo until it is experienced firsthand. Its study has been an area of interest since the early days of aviation and is a critical subject of investigation for specialists in aerospace medicine. Vertigo can affect a pilot's or astronaut's situational awareness, the accurate perception of the many factors affecting the flight of an aircraft or spacecraft, and thus has been studied extensively by aerospace physiologists. Most disorienting illusions experienced by pilots involve misperceptions of how they are oriented, due to a mismatch between signals from the balance apparatus of the vestibular system and visual cues associated with looking outside the aircraft.

The vestibular system is located in the inner ear and is made up of three semicircular canals that detect rotational movement in three different axes, and otoliths that enable us to perceive linear acceleration. The visual inputs from our eyes are integrated with the inputs received from the vestibular system and — based on our tumbling, rolling and dizzying experiences as a child — we use the combined information we receive from both to make conclusions about the world around us. For pilots and astronauts, however, those learned experiences can cause disorientation during critical phases of flight. One of the classic misperceptions experienced by pilots is the confusion between pitch and acceleration. Rapid acceleration can easily be misperceived as the backward pitching associated with a steep climb in an airplane. Similarly, rapid deceleration can be perceived as steep descent or even a forward tumbling sensation. When combined with matching visual cues, the misperception can be very compelling.

This perceived link between tilt and linear movement is routinely used in modern aircraft simulators to give pilots the sense that they are accelerating on the runway to takeoff and decelerating on landing. The simulator tilts dramatically to give the pilots inside the feeling that they are picking up or losing speed in a linear fashion. And in routine naval flight operations, it is not uncommon for the rapid acceleration of a nighttime catapult launch from an aircraft carrier to be misperceived as the aircraft pitching up. The natural response of the pilot would be to lower the nose, which could result in a catastrophic crash into the water. Fortunately, these highly experienced pilots have learned to rely on their instruments, which show the airplane rapidly accelerating, not acutely pitching up. Less experienced pilots not trained in instrument flight can be fooled and it is not uncommon for pilots taking off from isolated communities at night, where there are no visual cues from lights on the ground, to confuse the acceleration of the aircraft with a rapid climb and lower the nose, occasionally resulting in a tragic descent into the ground.

Training can improve the performance of pilots and astronauts in situations where vestibular and visual cues provide confusing inputs that could otherwise lead to a catastrophic outcome. But what happens if those sensory structures are themselves affected by prolonged exposure to microgravity during a spaceflight? Evidence shows that long-duration spaceflight can affect post-landing operator proficiency. Whether landing a space shuttle, flying an airplane or driving a car, the sensory changes associated with spaceflight affect performance.

Results from the space shuttle era demonstrate that even short-duration missions can affect the landing performance of space shuttle commanders. A team of researchers worked with NASA scientists to study this phenomenon, and upon reviewing existing landing data they found that the touchdown speeds for the first 100 shuttle missions demonstrated that 20 percent of orbiter landings were outside the acceptable limits, and the maximum main gear tire limit of 217 knots had been equalled or exceeded six times. The hardest touchdown in the program occurred following the commander's momentary loss of orientation, experienced after an abrupt head movement just prior to touchdown.[7] These findings led researchers to study the performance of ISS astronauts after their return to Earth in several simulated scenarios, including landing a high-performance jet, driving a car and driving a Martian rover.

Professor Steven Moore from Central Queensland University in Australia led the research project and found that driving was the biggest challenge. "It was quite interesting because a lot of them mentioned that during the six months in space they'd lost the sense of how big a car was around them. Essentially, we just had them driving along a 3-km [winding] mountain road and they had to just maintain lane control. They crossed over onto the wrong side of the road, they took longer to correct, they tended to hit the guardrail.

"The good news is . . . once they'd had a go and had problems, the second attempt in all cases was good," Professor Moore reported. "That suggests if you give them enough just-in-time refresher training towards the end of a long transit, that [their performance] should be recoverable."[8] The results suggest that for a successful landing on Mars, the first 24 to 48 hours after the six-month voyage will be critical for astronauts to readapt to working in a gravitational environment. But what about shorter trips to the Moon? Will astronauts have similar problems?

During the later Apollo lunar missions in the early 1970s, astronauts evaluated the driving characteristics of the lunar rover, providing NASA with their perspective on its handling capabilities. Did the Apollo astronauts have concerns about driving the lunar rover on the Moon? With NASA's sights set on future lunar landings, in 2008 a conference was held in Houston, Texas, to discuss the medical challenges associated with going back to the Moon. Participants were surprised to learn that although the top speed of the lunar rover was only around 11 mph, some of the Apollo crew members said rover operations posed the greatest risk of injury among all the lunar surface activities they undertook. They reported that during rover operations, they often misperceived the angles of sloped terrain, and bouncing into and out of smaller craters while travelling cross-slope at times caused a feeling of nearly overturning, which resulted in the crewmembers reducing their rover speed.[9] Based on the findings of Moore's team, it sounds like all that's required is a little lunar driving experience. Fortunately, there are no lanes or other traffic to worry about!

Understanding how the body adapts to space will help with future missions back to the Moon and may also give insights into the mechanisms underlying clinical conditions such as vertigo or osteoporosis on Earth. In addition to formal studies by space researchers, astronauts are also interested in the changes they notice when they fly in space. Astronaut Tim Peake shared his experience doing his own experiment to assess the role of visual and vestibular inputs in dizziness in space. "Something that happens when astronauts first come into space is that they usually feel pretty rough for about the first 24 hours, a mixture of dizziness and becoming disorientated . . . And I think a lot of this has to do with the fact that the vestibular system is a little bit messed up . . . that the brain sorts it out after about 24–48 hours and shuts down the vestibular system and relies more on information that's coming from the eyes." In his experience, "that helps you because it stops the dizziness, it stops the disorientation, there's no more nausea, and you can turn yourself in all sorts of orientations . . . upside down and (you can) spin about, and it doesn't really matter."[10]

Always interested in how he was adapting to spaceflight, Peake shared, "I wanted to do a quick check today . . . just to see if I can make myself dizzy, and if I can make myself dizzy, how quickly can I recover . . ." Enlisting the help of fellow astronaut Tim Kopra, he wanted to "get Tim to spin me around, doing something that would probably make me feel quite sick down on Earth." Smiling, he commented, "This could be the worst, worst idea I've ever had." Pulling his arms and legs into his chest, Peake had Kopra spin him at ever-increasing rates to see what would happen.

"Yeah, let's go for it. That's good. Okay, I'm just starting to feel the G's of acceleration there. That's really fast from my point of view. . ." Peake was seeing the world whizz by but wasn't feeling ill at all. He had no symptoms of nausea or any feelings of dizziness until he started moving his head while spinning. "[If] I've got my eyes open, when I move my head, that is definitely provocative. That gives me some off axis [rotation], but again it's not making me feel sick. That's quite weird." When Kopra brought him to a stop, Peake said, "I'm feeling dizzy." He waited to see how long it would take to stop and felt completely normal within a few minutes.

Space motion sickness was first described by both American and Russian astronauts, but no reports of it were made during the Mercury and Gemini programs in the early 1960s. The first episodes were reported later in that decade by the Apollo astronauts who were inside larger spacecraft and were able to get out of their seats and float around.[11] Around 30 percent of the Apollo astronauts had symptoms of motion sickness early in the mission but none while working on the lunar surface.

When NASA was getting ready for the Skylab space station program in 1973, space medicine expert Dr. Charles Berry and physiologist Dr. Jerry Homick speculated on the role artificial gravity might play in preventing motion sickness and some of the other changes associated with adapting to life in space. They concluded that virtually all astronauts are able to adapt in the first few days of the mission. The leading theory for the cause of space motion sickness was, and still is, a mismatch between visual and vestibular information similar to what happens in seasickness.[12] But other factors can contribute as well. Berry described the role of emotional factors when he reported that one of the Apollo 7 crew exhibited symptoms before he became airborne — perhaps not surprising as some landlubbers can get sick simply looking at a boat. There are likely several things at play in that strange experience, but anxiety may be partly to blame; the stress about possibly getting motion sick can be a factor in bringing it on.

In the space shuttle missions, 60 to 80 percent of astronauts experienced some degree of space motion sickness for the first two to four days of the mission, with symptoms ranging from decreased appetite and stomach awareness to vomiting. Most likely, the frequency may be linked to the increased mobility of the astronauts early in the mission. Often it starts while floating around while getting out of the spacesuit that's worn during liftoff. Although feeling queasy or even vomiting is generally not a problem on board the spacecraft, the risk of experiencing motion sickness during a spacewalk and choking inside the helmet led to scheduling these activities later in the mission, after the astronauts had adapted to being in space. The adaptation of the vestibular apparatus to microgravity has continued to interest researchers for its role in motion sickness and the many different illusions experienced by astronauts, both during and after spaceflight.

Many on the space shuttle have reported a sense of tumbling when the main engines are shut down and there is a rapid decrease in acceleration after the eight-and-a-half-minute ride to space. This phenomenon can be particularly entertaining after returning from space, when every head movement produces an exaggerated sensation. Simply moving the head forward then suddenly stopping the motion can create the illusion of tumbling forward. Backward movements of the head result in a sense of backward tumbling. Quickly turning the head to the side might result in a sense of whirling and dizziness like that experienced by children spinning around in a circle, and even small head tilts feel like a significant lean. It is almost as though the lack of gravity in space has increased the responsiveness of the vestibular apparatus in the absence of any strong gravitational stimuli.

Dr. Steven Highstein of the Washington University School of Medicine studied this phenomenon on the NASA Neurolab shuttle mission in 1998. He suggested that the microgravity of space would result in the gravity sensors in the ear becoming more sensitive to inputs, which might be detected by studying the activity in the vestibular nerve after a spaceflight. Despite the general willingness of astronauts to participate in research, this was an experiment better suited for different subjects.

Highstein and his team chose the oyster toadfish because its vestibular system is similar to mammals and it was easier to implant an electrode in the vestibular nerve of the fish to record its output in space and immediately after return to Earth. The day of landing, the responses to a side-to-side movement of the fish were on average three times greater than that of the non-spacefaring control group. It appeared that there was increased sensitivity in the gravity-sensing portion of the vestibular system after being in space. The time it took for the fish to return to normal was similar to the amount of time reported by astronauts for their disorientation and altered balance to improve following their return from space.[13]

Despite the progress in understanding how astronauts adapt to space and readapt on returning to Earth, optimizing these transitions will be important for future missions. Years ago, *The Far Side* cartoonist Gary Larson published a depiction of an alien spaceship landing on Earth. The drawing shows a group of unimpressed earthlings gathered on the ground, looking on as one of the aliens has just tumbled down the stairs of the spacecraft. The other aliens remain in a group at the top of the stairs and one chastises their clumsy colleague: "Wonderful! Just wonderful! . . . So much for instilling them with a sense of awe."[14]

Although we humans may not have to worry about trying to impress alien onlookers on the Moon or Mars, most astronauts would prefer to avoid that vertiginous feeling when they climb down from their spacecraft. Anyone who has experienced motion sickness or vertigo understands how significantly it affects daily activities. Some are virtually incapacitated by their symptoms as they struggle to perform the simplest tasks. Preventing the dizziness caused by spaceflight is an important priority for NASA to help astronauts succeed, and what we learn might also help clinicians develop solutions to manage similar problems on Earth. Working in the extreme, harsh environment of space, it is critical that astronauts are immediately able to focus on their objectives. They need to quickly achieve high levels of performance before they can take their first, historic steps on another planet.

CHAPTER 4
Food for Thought

I n the year 2374, Ferengi bartender Quark from *Star Trek: Deep Space Nine* had a problem: his drink replicator, an essential for any interplanetary pub, was not working.[1] Quark was grouchy because his emergency maintenance request was not going through the chain of command as quickly as he needed to keep his customers happy. Fortunately, it appears that all key *Star Trek* crew members have some ability to repair things. Dax, the ship's counselor — essentially a person who assists with everyone's mental health — happened to be on break and willingly dove into the replicator to pull out some green goo gumming up the works.

And think of the number of times Captain Jean-Luc Picard of *Star Trek: The Next Generation* began a meeting with the command "Tea. Earl Grey. Hot,"[2] and a mug of steaming drink would appear like magic under the replicator. Is space food this easy in real life?

In reality, the first iterations of space food were typically unpalatable enough to make you want to stay home. John Glenn was the first American to eat in space, during his orbital Friendship 7 mission in 1962. He ate applesauce (from a tube) and xylose sugar tablets with water. Not much of a meal but, to be fair, the flight lasted less than five hours and doctors back then were disputing whether Glenn and other astronauts would even be able to eat, swallow and properly digest food in weightlessness.[3] The Soviets, of course, had their cosmonauts eating as well; Yuri Gagarin was the first, chomping beef and liver paste from a tube in 1961, several months before Glenn's modest repast.[4]

After Glenn, Mercury astronauts on missions of a few hours survived eating Army survival rations — puréed food in aluminum tubes and similarly unpalatable items. Gemini astronauts on missions of up to two weeks ate bite-sized, freeze-dried or dehydrated foods. No wonder astronaut John Young of Gemini 3 famously smuggled a corned beef sandwich on rye onto the spacecraft, although the astronauts quickly stopped eating it — they were alarmed at the crumbs it was producing, which could clog the air intakes, leading to disastrous consequences.[5] Today's International Space Station crews substitute pita bread because it fulfills the carbohydrate requirements with far less mess and fuss.[6]

Thankfully, the food began to improve by the time Apollo astronauts were preparing for Moon landings in the late 1960s and early 1970s. Instead of injecting dehydrated food with cold water, the astronauts could use hot water if they chose, thanks to improvements in the spacecraft water system. Skylab astronauts had the option of refrigerating food, thanks to the roomier quarters of the space station.[7] Happily, the shuttle program had an even more expanded array of food that allowed the crew to have different meals for up to seven days in a row, and even to customize for different cultural requirements. The food was prepared in a galley on the mid-deck, using a water dispenser for rehydration and a galley oven to heat things up.[8]

Long-duration space living presents greater food challenges, however. Cosmonauts on the Mir space station were trying to pioneer feeding people for up to a year at a time, but the crews still mostly used rehydrated foods with the occasional supply ship bringing up fresh foods. That said, it appears the Soviet and Russian crews enjoyed some variety, as American astronaut Andrew Thomas complimented the selection of soups and juices after five months visiting Mir.[9]

On the ISS today, crew members can choose from roughly 200 different items on a standard menu, and food items can be customized to include some commercial items (beef jerky is a favorite) as long as they meet all the standard safety requirements (like not producing crumbs). But fresh food is still a rare thing, dependent as it is on supply ships; between deliveries, all food items must be shelf-stable.[10]

In space, where monotony and boredom are an issue, astronauts look forward to their meals to help break up the day, but keeping them psychologically interested is a challenge. We've all had the experience at home of looking in the cupboards and being dismayed that we have only one kind of soup available; imagine the additional challenges for people in remote rural areas who can only get to the grocery store once in a while. Multiply that logistics problem by a few factors and it is easy to understand the issues of shipping food up to a space station or spaceship.

Two generations of studying food in space is also revealing much about the gustatory challenges and food preferences of astronauts. The sense of taste is dull and smell is not as effective, which may lead to astronauts eating less than required. Still, our understanding of how digestion is affected in space is somewhat preliminary. As we consider month-long voyages to the Moon later in the 2020s and much longer journeys to Mars, perhaps in the 2030s, we need to ask ourselves: How do we provide enough variety and nutrition amid all of the other challenges of spaceflight?

To figure out what's happening to the body in space, it's helpful to consider a quick survey of the digestive system, from head to colon, while you're floating around on the space station. The cause of dulled taste and smell in space is the well-known "fluid shift" problem. Recall that in microgravity, water in your body tends to drift up to your head, producing the famous puffy-face syndrome and nasal congestion. Crews often ask for hot sauces and spicy foods, particularly in their first few days, to make everything taste a little more flavorful.[11] While the puffy face and fluid shift subside in a few days, more problems can arise when weeks are spent in space.

We now know for sure that astronauts can chew, swallow, digest and pass waste in space; it's been done time and time again on short-duration missions of a few hours, days or weeks. Astronauts commonly get questions about how to go to the bathroom in space, and the answer is "very carefully." The toilet — which in essence is like a vacuum cleaner, as it is designed to suck things away from you into the inner workings of Universal Waste Management system — enables astronauts to "boldly go" where few have gone before. As they say, "if you break it, you fix it," so learning how to use the toilet is critical! On Earth gravity is our friend and everything falls to where it needs to go for it to be flushed down the drain. In space, aim is critical and you don't want to miss the narrow opening of the toilet.

The airflow of the suction helps with accuracy, but stories abound of situations when things went awry. The Apollo astronauts had a more rudimentary system for disposing of solid waste. They had to use "Apollo bags" — special bags containing disinfectants that would be used to collect their waste. Once finished, the bag was sealed and the contents kneaded with disinfectant before being stowed in a waste receptacle.[12] In 2013, NASA released the mission logs of the Apollo 10 mission, which graphically documented the problem of a "turd floating through the air."[13] It's no wonder that some astronauts on short-duration missions have tried to hold off as long as they can, but inevitably nature calls.

To void there is a separate suction tube with a funnel attached at the end with different shapes and sizes available to accommodate both genders. Cleaning up takes time and starts with cleaning private parts, the toilet, hands and anything else that may have become contaminated. The final step is wrapping up any wet wipes or dry trash into a small tight-fitting ball of trash and stowing it with the bio-contaminated wet trash. It's generally at this point that most astronauts decide there's no way they're getting up in the middle of the night to go through this again.

A weirder effect of microgravity is that burping can be riskier than it is on Earth. Without gravity, gases don't float to the top of the stomach contents as they normally would. A typical burp provides relief from swallowed gas, often in a particularly sonorous manner. In space, acting upon the desire to burp might randomly result in either air coming up or . . . stomach contents. Fortunately, many astronauts don't feel the need to burp as frequently and avoid the risk of this random event. This intriguing detail shows just how much there is to learn about human digestion. In fact, experts are still trying to figure out the effects on digestion after extended stays in space. The rise of more advanced medical equipment that is portable and accurate makes these changes easier to track as the years go on.

In 2019, a study led by researchers at Northwestern University found that spaceflight affects bacteria diversity in astronaut digestive systems. Over the past decade researchers and clinicians have found that the microbial organisms found in the gut — its microbiome — are an important element in the prevention and management of a wide range of inflammatory and metabolic disease. The team at Northwestern, using a tool called the Similarity Test for Accordant and Reproducible Microbiome Abundance Patterns (STARMAPS for short), focused on a study of mice in space. Rodents are a popular substitute for human test subjects, but the researchers were also careful to compare the data set to previous missions on the space shuttle and International Space Station, as well as the Twins Study and research on Earth on the effect of radiation on the gut microbiome.[14]

Given the growing awareness of the importance of the human microbiome in maintaining health, NASA physicians were interested in understanding the implications of the Northwestern study for human spaceflight. The researchers found "an elevated microbiome alpha diversity and an altered microbial community structure" in mice after 37 days on the International Space Station — in other words, more microbes and a different set of microbes. Equating mice and human studies can be a stretch, but the early implication is that certain mammals may have a different interior microbe environment in their guts when exposed to microgravity for an extended period of time, which may in turn affect digestion and similar processes. The changes the researchers saw were comparable to the findings from a previous two-week space shuttle mission, STS-135, that also carried mice into space. The observations from both missions suggested changes in the chemical reactions in the mice that convert food to energy, and the researchers reported an altered expression of genes in the liver. What all this means for astronauts is the subject of ongoing research.

The effects of space radiation on the gut are still poorly understood, and the study's researchers have commented that either radiation, microgravity or both may alter its microbiome, "representing a risk to astronaut health, especially during long-term spaceflight missions." As studies continue to evaluate the role of radiation and changes in gastrointestinal physiology induced by microgravity, other research suggests that microbial physiology and growth may change due to a variety of factors such as inflammation, the infamous fluid shift, cardiovascular changes from spaceflight and altered physiological responses in digestive function.

It may also be that the microbiome of the gut alters in space because the crew members are in an enclosed environment and working in relatively close quarters, ultimately sharing a common bacterial makeup. Another 2019 study, this time of nine long-term space explorers who spent six to 12 months on ISS, showed "composition of the intestinal microbiota became more similar."[15] Again, the small sample size makes it difficult to draw conclusions.

One metric among many that doctors use to assess digestive health is "gastrointestinal transit time." Digestion time varies widely between people even on Earth, but it typically takes about 36 hours to get food through the entire colon, according to the Mayo Clinic.[16] In this time, the food spends about six to eight hours moving through the stomach and small intestine before passing into the large intestine or colon for digestion and water absorption. Eventually, two to five days after eating, the undigested remainder is passed as feces.

One medical study from STS-51, a shuttle mission that lasted 10 days in 1993, suggested that gastrointenstinal transit (GI) time might increase during spaceflight — although the sample size was quite tiny. The researchers studied two astronauts at various points: 60, 45, and 30 days before launch, the fifth day of flight (about halfway through) and nine days after landing, by which point the astronauts would have been close to fully adapting to the rigors of Earth once again.[17]

At each collection point in the experiment, the astronauts were given an oral dose of the drug acetaminophen (the active ingredient in Tylenol) because, the researchers noted, "The absorption rate of this drug after an oral dose is directly proportional to the gastric emptying rate." Then, they collected saliva and urine from the subjects, including metabolites in the latter case. The GI transit time was measured by subtracting the "gastric emptying time" of acetaminophen from the transit time observed with a breath test. Unfortunately, the researchers determined that the sample size of astronauts was too small to draw a meaningful conclusion, although they note the transit time appears to increase during flight.

It appears there has been little other work done in space to compare with the results of this experiment. A lengthy 2014 NASA research report on nutrition in space, with data collected from three space nutritionists, noted that "gastrointestinal transit time has not been systematically studied during flight." However, a bed-rest study on Earth suggested that the mouth-to-cecum timing is "significantly longer" when the subjects are in the standard space position, slightly head-down, for 10 days. If GI transit time is indeed longer in space, it's unclear what the effects may be. Anecdotal data from

Skylab and other missions, the researchers said, suggests that sometimes astronauts may consume less food. Yet this is likely not tied to GI transit time; rather, the researchers suggest it might be related to space motion sickness, as usually the lesser food intake persists only for the first week or two at most, when astronauts are still adjusting to spaceflight.[18]

Another problem associated with time in space is constipation. NASA is still trying to understand exactly why that happens, although there are some likely factors. Bed-rest studies suggest that the lesser movement in bedbound people (which is akin to space) may induce a condition called hypokinesia, which can produce problems such as loss of appetite, indigestion and constipation. And that's not all: "This suppression in appetite will lead to negative nutrient balance thus causing a change in energy level and cognitive function," warns a NASA study focusing on bed-rest subjects, adding that decreased energy "in turn decreases motivation and cognitive ability."[19]

While constipation in the first few days or weeks of a mission could be attributed to hypokinesis associated with microgravity, fluid shifts or other factors, astronauts on longer-duration missions have also reported other ongoing issues, suggesting the fluid shift problem may persist or that other bodily systems, like the cardiovascular system, may be contributing to long-term digestive issues. Researchers have expressed concerns that in the long run, this might result in decreased overall immunity in some astronauts,[20] but as most of our work in this area has been done with bed-rest studies and rodents, it is hard to say for sure.

Gastrointestinal changes may also alter ways in which the body absorbs vitamins, the NASA report noted. This could account for a potpourri of difficulties that astronauts have historically encountered in space, and that bed-rest analog astronauts experienced on the ground. While the small crews and the relative great health of each person in space make it difficult to confirm all this, we'll surely get a lot more data from future missions on the International Space Station and perhaps private spaceflight as well.

Let's run through a few of the difficulties the body may experience because of alterations in gastrointestinal functions. It's long been known that bone formation is an issue in space, although it appears that countermeasures such as NASA's Advanced Resistive Exercise Device do slow or stop these problems. Historically, studies found that a handful of Soviet or

Russian Mir astronauts — who had no access to this high-powered device — had increased resorption of bone and that bone formation continued to be slowed for two or three months after landing from a long-duration mission. Scientists are still struggling to understand why this happens, but it appears digestion is a factor. As calcium is released from the bone, researchers noted in quoting other studies, there are lower levels in activated vitamin D, "which leads to a reduction in calcium absorption from the gastrointestinal flight." The answer isn't as simple as popping calcium supplements, which is why resistance exercise is seen as the preferred countermeasure for the time being.[21]

Iron sufficiency also seems to be affected in space, but again taking supplements may not be the answer. Researchers agree that at close to toxic levels, iron could be a leading factor in cancer or tissue damage. Consistent abnormally high levels of iron also seem to be related to "gastrointestinal distress," the researchers said, although they did not get into detail. Notably, consuming certain foods can lead to lower iron absorption, such as tea, coffee, bran, calcium, phosphate, egg yolk, polyphenols and some dietary fibers, which is why it's so important for astronauts to consult nutritionists and follow their menus in space.[22]

There's also a worry about fluid intake. Doctors get concerned about astronauts who are ill over long periods of time, since diarrhea and vomiting can lead to dehydration when losses occur in the gastrointestinal tract.[23] Researchers emphasize the importance of maintaining adequate fluid intake; interestingly enough, it appears that astronauts (left unguided) may take less fluids in space than they do on the ground, and the amount they consume is "below the recommended quantity."[24] This is why it's so important for crews to have access to fresh water. On the ISS, a novel recycling system allows for safe recycling of urine into water,[25] which decreases (but does not eliminate) the need for shipments from Earth. Fresh water is an essential for healthy living in space, so this study will be helpful to see how well recycling works in the long term and what changes need to be made. In the short term, however, water resupply from Earth is available in case of failure on the ISS. It's a highly complex system, but in simple terms: After an astronaut uses the toilet, a rotating distillation unit separates liquids and gases. Filters and a processor remove unnecessary gas and solids, while microbes are zapped using a high-temperature chemical process.

So far, we've been focusing on the effects of microgravity on the gastro-intestinal tract. Certainly, these findings will be useful during long-term stays on the ISS and during the long voyage between Earth and Mars, but can you also have trouble with your digestive system in low-gravity environments such as the Moon, which has one-sixth the gravity of Earth, or on Mars, where gravity is one-quarter of Earth's?

Astronaut John Young certainly thought so. Young was commander of the ambitious 12-day Apollo 16 mission in 1972, which touched down in the Descartes crater of the lunar highlands. Geologists were on the hunt for younger rocks on the Moon to learn about signs of change in our neighbor's lengthy 4.5-billion-year history. The astronauts were equipped with geology training, numerous tools and a lunar rover to work on the surface.

One night after working on the lunar surface, the astronauts began talking to each other about their orange juice intake. The crew accidentally had a setting on their microphones activated, called a "hot mic," which automatically transmitted their voices to Earth. The following conversation[26] was therefore broadcast to NASA in real time — and, presumably, heard by a significant number of people who were following every move of the second-to-last landing on the Moon.

"I have the farts, again. I got them again, Charlie," Young said to his crewmate, Charlie Duke. "I don't know what the hell gives them to me. . . . I think it's acid stomach. I really do."

Duke assented: "It probably is."

Young continued his rant. "I haven't eaten this much citrus fruit in 20 years! And I'll tell you one thing, in another 12 . . . days, I ain't never eating any more. And if they offer to [supplement] me potassium with my break-fast, I'm going to throw up!"

He paused for a moment, then continued. "I like an occasional orange. Really do." He laughed. "But I'll be darned if I'm going to be buried in oranges." Young was from California, a state renowned for its citrus fruit, so it's probable he'd had a lot of the fresh stuff growing up, but they seemed to affect his gut much differently in space.

The doctors attempted to correct concerns about the risk of abnormal heart rhythms with low levels of supplemental potassium on Apollo 16, but only with partial success. A *New York Times* report of the era said the 16 crew had to double their potassium intake, but their bodies did not register double potassium levels as hoped. Also, the newspaper laconically pointed out, "the astronauts did not like it" due to the digestive issues.[27]

The true answer to figuring out the digestive issues related to space travel is to take a big step back and admit that we know practically nothing. A NASA publication, speaking about the effects of gravity on the nervous system, admits something that is just as true about almost any other system in the body: "We still know little about the impacts of long-duration microgravity exposures, the effects of partial gravity environments . . . and how to develop effective physiological countermeasures."[28] Today's astronauts are much more used to having experiments performed on them, so it would be no surprise if future astronauts in the Artemis program would be more wired-up while on the Moon than their Apollo predecessors were, to track changes in their bodies — including changes to the gastrointestinal system.

While NASA and other agencies consider the challenge of human health and digestion, the agency is thinking hard about how to make food more palatable and available on other worlds. One result was the Deep Space Food Challenge, announced by NASA together with the Canadian Space Agency in 2021. The goal, according to the program's website, is to "create novel and game-changing food technologies or systems that require minimal inputs and maximize safe, nutritious, and palatable food outputs for long-duration space missions, and which have potential to benefit people on Earth."

"Minimal inputs" is important because food preparation should not be too taxing on an astronaut's time, especially during the workday. Many of us on Earth attempt to get organized for the week by preparing meals ahead of time and then storing them in the fridge or freezer to reheat quickly during the week. Refrigeration is not typically an option in space, and astronauts similarly don't want to spend a lot of their workday making meals when there are other tasks to accomplish.

In space, the "outputs" — the food that you produce for the time you put into it — should ideally be nutritious and easy to prepare, like when you use the microwave as a quick and convenient way to cook or warm food. Time is at a premium in space, where care must be taken to accomplish everything in an inherently dangerous environment, and where you constantly need to deal with time-consuming tasks like donning a full spacesuit just to catch a bit of sunlight outside on the Moon or Mars.

Moreover, any food devices need to fit strict physical dimensions as well as mass, power and water requirements so as not to overtax the space station or other facilities where astronauts are staying.[29] Also, it's essential that your outputs don't include "hazardous compounds or materials" that might include microbes, bad gases or other "toxic components," the NASA documents urge. This includes avoiding any "physical, chemical or biological hazards associated with the hardware or the process."[30]

The contest is still at an early stage, with the winning selections to be determined in the coming years. We can't easily predict what devices will be developed or what creative new food choices will look like, except to know that the objective is to output food products that the crew will enjoy and that will be safe for space. Naturally, another big goal is to figure out how to grow plants on the Moon or Mars, following studies from the ISS and the shuttle missions. For those optimistic that the Red Planet could yield potatoes like in the movie *The Martian*, preliminary studies do show some hope. One 2014 effort, for example, found that plants could survive for up to 50 days in simulated lunar or Martian regolith (the dust or rocky, soil-like layer that covers the bedrock) with no nutrients added.[31] That said, we'd need to stretch the capabilities many more times for a long Mars mission.

It's fun to zoom out to the big scale and ask how a future Martian settlement could safely feed something like one million people. Surprisingly, a 2019 study in the journal *New Space* suggested that such a population could attain self-sufficiency in only 100 years.[32] On this massive scale, you must think locally to produce as much food as possible. Solar power and fission reactors would be essential for energy. Water could come from the ice caps and, potentially, from resources infused within the regolith or under the surface. NASA also plans to study the conversion of carbon dioxide into oxygen on a small scale on the Perseverance rover,[33] which could be useful to generate the vital gas on a larger scale in the future.

Some of the food solutions to support colonies on Mars may surprise you. Insects such as crickets are a highly nutritious source of food — they're high in caloric value and require relatively small amounts of water and nutrients to keep healthy, supporting the idea of minimum inputs for maximum output. Insects as a food source are not widely accepted in most European or North American societies, but perhaps a little seasoning would make them more palatable to more people.[34]

Other options might come from what the researchers term "cellular agriculture," meaning food grown from cells embedded in lab dishes: think eggs (without the chicken), milk (without the cow) and even fish. If you prefer more naturally grown food, plants such as peanuts, wheat, soybeans, corn and sweet potatoes are a possibility as long as you use strong LEDs supplemented with sunlight.

Of course, we're nowhere near figuring out the complete solution to feed this hypothetical colony, but there's still time to pursue a few directions, Kevin Cannon and his team of scientists from the University of Central Florida told Space.com in 2019. "Future research on how to best feed Martians should focus on boosting crop productivity, working out the most efficient and palatable insect species, improving the flavor and texture of cultured meat, improving the efficiency of the LED lighting used to grow crops, and developing automated methods for rapidly building pressurized, shielded areas to house farms," the report said.[35]

Keeping astronauts eating healthy food on other worlds will be a challenge, but the food scientists are on it and there is optimism that astronauts will have palatable options to look forward to, while protecting their digestive tract — whether that means avoiding the farts from orange juice, finding a way to convince astronauts to eat insects, or thinking about what countermeasures may address some of the common issues with the gastrointestinal tract. The big lesson to remember about voyaging farther into space, just like any trip to a foreign country, is that it's essential to consider the local environment and the human body very carefully before doing anything too risky. North Americans are already well accustomed to taking precautions in terms of water intake and types of food when voyaging to places that have pathogens in the water that they aren't used to. Accordingly, we need to remember that the human digestive system faces even greater disruption from factors such as radiation, microgravity and other stresses of being in space — and that we should think through all countermeasures before going too far into the universe, just in case.

The many nuances of food — personal preferences, its availability and ease of preparation in space — are in many ways reminiscent of the challenges faced by past explorers. The ever-present risk of limited supplies, the need for a balanced diet with fresh fruits and vegetables, and the multicultural food preferences that enrich the mission are all factors to be considered when planning a lengthy voyage to another planet in our solar system. There is, however, one aspect of nutrition, food and the physiology of digestion that has been a constant throughout history — a great chef is critical to mission success! Perhaps there's an opportunity for a new reality TV series — *Chopped: A Space Odyssey.*

CHAPTER 5

The Space-Time Continuum

When the windmill-shaped Skylab station launched on May 14, 1973, NASA was looking forward to a new phase of human exploration. For more than 10 years, astronauts had hustled through short missions in low Earth orbit and on the Moon, remaining in space for no longer than a couple of weeks. With Skylab, the agency hoped to gather more detail on how the human body reacts during longer stays in orbit as successive crews of astronauts visited the space station on missions lasting a few months at a time.

Unfortunately, telemetry sent to the ground (Skylab was launched autonomously for later human occupation) showed that a micrometeoroid shield, which was supposed to protect the station from tiny space rocks and the Sun, accidentally deployed while still in the Earth's atmosphere. Skylab safely reached orbit, but when its solar arrays were extended, more telemetry showed the power generated was only 25 watts — far less than predicted — eventually leading to the conclusion that one of the solar arrays jammed during launch. Worse, temperatures in the unprotected station began to climb above what a crew could tolerate.[1]

The first crew, known as Skylab 2, arrived on the craft 11 days later and did urgent repair operations during two tricky spacewalks, freeing the stuck array and deploying a sun shield to cool down the station.[2] Skylab 1 was now ready to host crews. But as it turned out, NASA and the astronauts themselves still had much to learn about the human body and what kind of schedule is appropriate for long-term spaceflyers.

The Skylab 2 team spent a relatively short 28 days on the station, most of them action-packed with lots of time spent reconfiguring systems and working on interior repairs. Still, their stay doubled the previous record for time spent in space. The successor crew, Skylab 3, had more of a chance to settle into a long-term work routine with a nearly two-month (59-day) duration. In both cases, it seems that the crews had jam-packed schedules, which worked because the missions were relatively short; the astronauts could work at an intense pace. Skylab 4, however, would last 84 days and pose different challenges.

As with Skylab 2, Skylab 3 and many missions before, ground control would send a schedule up to the Skylab 4 astronauts every day for them to follow closely. As with past missions, that schedule had little room for error or adjustment because the perception — rightly — is that time in space is expensive and precious. Managing time is critical for all of us, particularly when we are under stress and trying to cope on our own, without the full support network we are used to. Think about what happened to so many workers who were forced to stay home during the global coronavirus pandemic. Adrenaline fueled us at first, allowing us to keep up with a demanding work schedule on top of constantly changing health guidance and, for some, juggling childcare and homeschooling. But after a few weeks or months in relative isolation, many people felt they needed a chance to slow down and adapt to the new reality. The same thing began to happen on Skylab 4.

In a NASA oral history recorded in 2000,[3] Skylab 4 commander Gerald P. Carr recalled what happened to his three-person crew: "The schedule was so dense that if you missed something or if you made a mistake and had to go back and do it again, or if you were slow doing something, you'd end up racing the clock and making mistakes, screwing up an experiment or not doing a procedure correctly."

Carr said the crew found this problem piling up for "many days," and although astronauts are always a motivated bunch, they found their morale dragging as they realized they could not get their assigned experiments done properly. The crew would work until 9:00 at night to try to catch up, which made it impossible for them to wind down for a planned sleep period at 10:00 p.m. Now sleep deprivation was a factor, on top of the lack of leisure time, slowing down the crew even more.

NASA then tried to make up the time by moving the crew's exercise period, sometimes scheduling it right after a meal. "That's no time to be exercising, particularly up there when you can't belch," Carr said. "With food floating [in your stomach], you're liable to get it back with your belch. So we started grousing at them about that, and they were working hard trying to keep up with the schedule, and we were giving them a hard time and they were giving us a hard time."

NASA and the Skylab 4 crew had previously agreed to 10-day workweeks followed by one day off, but the crew — feeling guilty at their testiness and at falling behind — began agreeing to take on tasks even when they were supposed to be resting. After a few cycles of this, the crew felt enough was enough and asked for a day off. NASA gladly granted it to them, but then came an incident that has been — in Carr's mind — misinterpreted in the half-century since the mission.

The crew was now 42 days into their mission, roughly at the halfway mark. It was December 28 and they spent their rest day using the station shower (Skylab was the last spacecraft to include a shower), reading and looking out the window, for the most part. In their tiredness, however, they forgot to get the radio ready for one of their planned passes over a ground station to chat with mission control. With radio silence coming from the crew after a tough six weeks, the press jumped to conclusions and said the crew was beginning to mutiny, explained Carr. "They said, 'Look at that. These testy old crabby astronauts up there won't even answer the radio now. They've turned off their radio and they won't listen to the people on the ground.' So we have lived under that stigma all these years, but basically it was we just got careless and we were busy doing other things and didn't think to configure our radios."

The crew tried jumping back into their busy schedule after this break, but found by about Day 50 that they couldn't keep up the pace anymore. Carr said they had a "séance" with the flight surgeon to show some of their readings, such as how much food they had eaten, while reporting that it appeared nobody — astronauts or ground crew — was happy with the schedule. Carr called for a "frank discussion," under the expectation of medical privacy, but the press somehow heard about it and "raised Cain," he said. With the press bothering NASA, ground control asked the astronauts to speak over an open channel so everyone could hear what was happening. The astronauts agreed.

Finishing the discussion took two orbital passes, with the crew giving their concerns during one pass — the need for rest, exercise, a lesser schedule — and ground control responding during the next pass with their concerns about a lack of flexibility in scheduling. The next morning, ground control sent a teletype message with suggestions, such as allowing astronauts to do small "housekeeping chores" on their own schedule, rather than when the ground thought was best. The ground also pledged not to bother the crew during their meals and to allow them time off after dinner.

"That really solved the problem," Carr said, adding that the new arrangement "worked beautifully" for the rest of the mission. Like Henry Ford's workers on the car assembly line 60 years before, Carr observed, the crew and ground found out that you can only work people so hard before they begin to make mistakes.

Best yet, the crew found the idle time allowed for more creative thinking, helping them to line up experiments in a way that made their schedules even more efficient. "Bill Pogue and I recommended very, very strongly to the International Space Station program that that's the way they [should] program their days," Carr added. "I think they took it to heart."

Unfortunately, the myth of a crew mutiny has persisted through the decades over strenuous objections by NASA[4] and the crew members themselves. One of the first reports was a 1976 article in *The New Yorker*,[5] and media still mention it to this day; even after all of the members of the crew had passed on, Pogue's 2014 obituary in *the New York Times* carried the headline "William Pogue, Astronaut Who Staged a Strike in Space, Dies at 84."[6]

Even though the Skylab 4 astronauts didn't actually mutiny, workload clearly had been a problem in the first half of their tour on the station. NASA hadn't had experience with any long-duration missions before this, and the agency quickly learned the importance of pacing and managing the astronauts' workload. Building on that knowledge, the longer-duration missions of the Shuttle-Mir program in the 1990s then informed planning for the International Space Station, which began accepting crews in 2000 for stays of a few months. Today, astronauts have much more flexibility in deciding how to manage their time, unless, of course, an emergency intervenes.

Managing time effectively can be tricky business. Think about the last time you pulled an all-nighter to get something done. Or that moment at the end of last week's final workday when you decided to squeeze out one last task, no matter how tired you felt. We make these little kinds of life choices all the time in the name of being more productive, but often this ends up backfiring. That's because time management is not so much about being efficient with your time, but being efficient with your energy.

"Most of us respond to rising demands in the workplace by putting in longer hours, which inevitably take a toll on us physically, mentally, and emotionally," the *Harvard Business Review* noted in 2007. "That leads to declining levels of engagement, increasing levels of distraction, high turn-over rates, and soaring medical costs among employees."[7] And that was even before smartphones blurred the boundaries between work and home. The global COVID-19 pandemic made that increasingly fuzzy line even worse when many office workers couldn't commute to closed-down offices.

World-changing pandemics aside, astronauts, workers and students today should realize that humans need maintenance, just as machines do. But it's not just about saying no to toxic work environments — as important as the notion is. Rather, energy management and life mainte-nance mean recognizing what your body needs. All living creatures need adequate sleep, a balanced diet, exercise, social activities and leisure time to thrive. It doesn't matter whether your job is chief financial officer in a high-rise office in New York City, a janitor mopping the floors in the same building during the coronavirus pandemic, or an astronaut floating in the International Space Station 250 miles above; regardless of your social status or that pressing work deadline, your body regularly needs chances to rest and replenish.

Launching to space is a stressful experience in itself, as the astronauts on board a spacecraft undergo several G-forces (one G-force is equivalent to Earth's gravity) for several minutes before being exposed to weightlessness when the rocket engines shut off. As we've seen, it's typical for astronauts to feel mild sickness and disorientation during their first few days in orbit, usually managed by mild medication and rest. NASA now also makes accommodations for its crew members to have a lighter schedule in this time — a valuable takeaway from the Skylab 4 mission. Spacewalking also does not take place within the first few days of a mission, a lesson learned from Apollo 9 when the first spacewalk by Rusty Schweickart had to be cancelled due to "space adaptation syndrome."[8]

NASA now has decades of accumulated experience — not to mention access to records from its international partners — to predict how the human body will react during those first days in space. That experience has led to NASA's ongoing "Spaceflight Standard Measures" program, which "characterize[s] how the body and mind change in space."[9]

The crew's adaptations in orbit are tested against analog astronauts on the ground in facilities around the world. The participants in the study take part in numerous experiments, such as wearing wrist monitors to measure activity levels and light exposure patterns — key to understanding sleep — and submitting to ultrasound scans to measure the thickness of the carotid artery in the neck, which sends blood to the brain; these scans will help researchers understand why arteries may stiffen in space.[10]

Ongoing studies in the space station environment also assess how to help crew members manage their physical and emotional energy levels. Sleep is essential to staying healthy and is an important part of managing crew energy, yet it is easily disrupted by noise and light. The ISS crew see a sunrise and sunset every 45 minutes, which might disrupt sleep unless window shades or eye shades are used. Recognizing the importance of ambient light in maintaining the body's internal clock, NASA and its partners sent adjustable light-emitting diodes to the space station in August 2016[11] to create better lighting than the existing fluorescents.

The space station has a small sleeping compartment for each of the crew, while astronauts on the space shuttle simply attached their sleep restraint system — also known as a sleeping bag — to whatever surface they wanted. For some, it takes a while to get used to sleeping while floating inside a sleeping bag, even if it's gently resting against a surface. There are no pressure points like there are when lying in bed and the experience gives new meaning to the phrase "floating off to sleep."

Most find sleeping in space extremely relaxing, awakening without the stiffness that sometimes accompanies a terrestrial snooze. In space it is amusing to find that over time some dreams are terrestrial in their content while others are based on being in space. Some retired astronauts have relived their spaceflight experiences years later when dreaming about floating in space as they lie in bed on Earth.

Rested or not, astronauts always have something to do. Sticking to the timeline is critical for mission success, so there's no sleeping in. Wakeup call is the start of the on-orbit day that many feel is so busy it's like a continuous sprint. It begins just like at home — everyone's figuring out what to have for breakfast. Dehydrated scrambled eggs are a great way to start the day, provided they're rehydrated for 10 to 15 minutes with hot water. Anything less and there might be powdered egg that floats out of the package when it gets opened.

The two most important things to have available are scissors to open the packages of food and a spoon to eat with. Forks and knives are available but almost everything can be eaten with a spoon. Breakfast tacos are fun to make (there's generally no bread in space due to its short stowage life). You can clip a taco to the front of a nearby rack of hardware, spoon the eggs onto the taco — the liquid in the eggs holds them to the shell — then add a little hot sauce, wrap it up and enjoy. Cleanup is easy — stow your trash, clean your spoon and scissors and then it's back to the timeline for the day's activities.

NASA sends daily instructions to the crew called the "Onboard Short-Term Plan Viewer," somewhat like an electronic day planner. Astronaut Scott Kelly, who spent nearly a year in space in 2015–16, characterized the viewer as a "relentless red line"[12] that only really ceases during the weekends, so it's not always welcomed by the crews. While many try to think of it as a guidepost for the day, similar to how most of us operate with our smartphones, the unrelenting list of tasks can easily become a source of stress for the crew.

Scheduling a typical three- to six-person crew aboard a space station is no easy task. The space station is relatively roomy, with several modules in which to spread out, but each crew member needs time in the bathroom for washing up, and each needs time on the various exercise devices to stay fit. Those readers with teenagers at home will understand the challenge of managing time in the bathroom, and simply moving between modules at the wrong time to engage in various activities could disrupt a press conference or disturb a scientific experiment.

Making things even more complicated, the NASA and Roscosmos (its Russian equivalent) mission control teams tend to operate independently, with each agency managing its own crew. It takes constant coordination and communication between the agency partners to make sure that everyone is doing the right thing at the right time.

"Fitting all the astronauts' activities into the graph isn't always easy," NASA wrote in 2004[13] about the planning tool. "Many variables have to be considered when laying out the schedule. This includes times that the vehicle is in the dark or light and the various time zones for the ISS, for Mission Control in Houston, and in Russia."

Each astronaut's workday is from about 6 a.m. to 9:30 p.m. Greenwich mean time (same as London; that's four or five hours ahead of New York City's eastern standard time, and five or six hours ahead of mission control in Houston, depending on when the clocks change between standard time and daylight saving time on either side of the Atlantic). That sounds like a long day, but it includes personal time before and after sleep and about 2.5 hours of physical fitness time, not to mention some leisure after dinner.[14]

A real-life morning routine from the first two hours of a day in 2014[15] shows the challenge in scheduling three NASA crew members in a space station on November 12, the day after Veterans Day, which is a federal holiday in the United States (managing statutory holidays in space adds yet another wrinkle due to the varying observances among different cultures). On a regular day, the commander is up at 6 a.m. space station time reading a daily schedule reminder, while the two flight engineers perform standard morning inspections and a laptop reboot. Each crew member is allocated a few minutes for "post-sleep" activities along with performing blood sampling for research purposes.

No, you haven't had breakfast yet — that's at 6:50 a.m. for flight engineers and 7:05 a.m. for the commander. Breakfast requires a lot of prep time and cleanup time in microgravity, plus NASA wants to give a little time for socialization, so you get about 50 minutes. That means the commander won't have time to finish breakfast before the daily planning conference with the ground commences at 7:40. "This is just a short 15-minute chat with the control centers in Houston, Moscow, and Huntsville (where our payload control center is located) to review the plan for the day and answer any questions we or the ground might have. Then we get started on the day's work," wrote astronaut Ed Lu, an Expedition 9 astronaut.[16]

By 7:55, the commander is squeezing in the last 15 minutes of breakfast while the flight engineers are working on their first experiment. Then everyone will be busy with their own activities until lunch at 1 p.m., at which the commander is expected to arrive five minutes late according to the timeline. That's how precise you sometimes have to be in space to get things done, with well-coordinated teamwork enabling everyone to work together in a tight environment.

As intimidating as the timeline may seem to an outsider, Lu noted that the astronauts are encouraged to jump in and make changes as long as they meet the goals.[17] "The ground doesn't micromanage our time, and in fact most things on our schedule are very flexible," he said.

"Only sometimes does a particular task need to be done at a particular time, and if so it is called out that way on our schedule," Lu added; typical examples might be a scientific experiment or maintenance task that requires the astronaut to work with a technician on Earth. Or, the crew member might be in a live event such as a press conference, a chat with schoolchildren or a call to the ground to speak with some dignitary or to celebrate a big event in space.

"Otherwise, we are free to move tasks around during the day as we see fit, although the order that Mission Control lays it all out in the schedule usually works pretty well," Lu said. "Mission Control's job is to see to it that the required tasks are prioritized and that there is enough time to do the tasks, as well as doing any needed coordination. Our job is to get all the tasks done. We often make suggestions for optimizing things, and we work together to make operations more efficient the next time."

NASA also has room in the schedule to deal with small to moderate problems. Although the agency does allocate time for routine maintenance, things inevitably break in space; so if a crew member is working on a broken toilet, for example, more room for maintenance will be added to the timeline. Sometimes the station also faces a minor "contingency," meaning that there is an issue that needs to be addressed quickly but not immediately. One of the more notorious examples is an ammonia leak. Ammonia is a vital coolant on the station for experiments and hardware, and at times astronauts have had to do spacewalks on fairly short notice (a day or two) to deal with the situation.

NASA also encourages its astronauts to have a lighter schedule on weekends, which Kelly said are often filled with different activities:

"videoconference with family, catch up with personal email, read."[18] Crew members also love looking out the window at Earth, particularly through the 360-degree Cupola window that is around the corner from some of the exercise machines. Speaking of exercise, the astronauts aren't allowed to take a break from that on weekends. Their bodies deteriorate too quickly otherwise, Kelly says, so that's about two hours of the day that must be taken up with personal maintenance.

The astronauts also do what many of us do on weekends — clean house. While most of us on Earth deal with cat litter and children's toys on the ground, astronauts clean in three dimensions because microgravity is fairly unforgiving in that respect. "On the space station, a piece of dirt can wind up on the wall, the ceiling, or attached to an expensive piece of equipment," Kelly points out. "A lot of crap winds up on the filters of the ventilation system, and when too much of it starts to build up, our air circulation is affected."

A more serious issue is dealing with mold on the walls, which tend to get "dirty and wet" over time, Kelly says. Luckily, cleaning the mold is not much worse than your typical bathroom-wiping routine; the astronauts use a vacuum and antiseptic wipes to remove the spores, and occasionally take samples to send back to Earth for the inevitable scientific experiments.

Sometimes, successful time management during a mission comes down to learning about human dynamics. Boredom and isolation are other common problems during long periods of time in space, and worse, astronauts do not fully control their own schedules. For example, you can't easily go outside if you want a change of pace; spacewalking is a highly regulated ballet dance that only takes place when strictly necessary, and not a joyride. And not all astronauts get to do extravehicular activities. So it's possible for astronauts to feel a little stagnant and confined after weeks or months of being in the same few rooms in space.

Just as conditioning specialists are assigned to the astronauts over a long period of time to ensure their physical health, NASA also has a behavioral health unit[19] devoted to the astronauts' mental and emotional well-being. This unit aims to help astronauts and their families through a typical training, deployment and post-deployment cycle; this means that these people will be working closely with an astronaut and their loved ones for three or four years. The agency tries to help astronauts as best as it can through training for "high-risk environments" — a little space station is just one bad hole away from a big problem — and managing any issues about work-rest schedules, fatigue or simply missing loved ones at home as the mission continues.

For most long-duration astronauts, at some point the novelty of the microgravity experience wears off and the effects of living in isolation begin to emerge. Many have described a typical six-month mission as a sprint with innumerable tasks that keep them busy and help distract them from missing friends and family at home. Along with that, though, boredom can arise from the isolation and many behaviorists studying the experience of long-duration astronauts understood when some described the monotony that can come from doing the same thing, in the same environment, for six months or more. NASA has been very supportive of crew requests for items that can help amuse them: crew can watch movies, read favorite books, play musical instruments, photograph the Earth and play different micro-gravity games (how far away can you be for someone to successfully shoot an M&M candy into your mouth?).

Astronaut Mark Kelly, Scott Kelly's identical twin brother, upped the ante when he sent a surprise package containing a gorilla suit to Scott during his one-year spaceflight. Why, you might wonder? "Because there's never been a gorilla in space before," Mark told Scott. The suit was a hit on social media in February 2016 when Scott, dressed in the gorilla suit, emerged from a big white bag to chase British astronaut Tim Peake around the station.

It seems less likely that such antics would have taken place in the past, but even the Apollo astronauts had time for a bit of fun on the Moon. Apollo 14 astronaut Alan Shepard took a few minutes during a live broad-cast from the lunar surface on February 6, 1971, to hit two golf balls with a specially crafted lunar six iron, commenting that the second shot kept going for "miles and miles."

Variety can also come through changing the routine, especially when NASA sends resupply ships to the ISS. The astronauts invariably look forward to the little surprises in their "crew care packages," sometimes put together by family, or sometimes by support staff to help enhance the envi-ronment just a little. "Maybe it's a picture drawn by their child or a favorite snack; no matter the item, those extra touches mean a lot when isolated from home," NASA says. These packages also may include fresh fruit and vegetables, which are rare treats on the station.

What cannot be overstated is how much NASA builds crew integration into the training and the space missions themselves, which saves time on mission because everyone knows each other's habits. One of the reasons a typical six-month crew trains for two years is to really get to know one another, to the extent that (to paraphrase the 1995 movie *Apollo 13*) such a lengthy training period allows individuals to get to know each other through the tones of their voices. All crew members have a working knowledge of Russian and English, allowing conversation to flex and adapt to support crew members from different countries. If a problem comes up or if frustration arises, crew members are taught from their first day of astronaut candidacy school — and even during selection, in some cases — to rely on each other to solve problems.

In the case of crew dynamics, studies are ongoing at NASA and at analogs, environments on Earth that try to simulate situations in space. One famous example is the HI-SEAS (Hawaii Space Exploration Analog and Simulation) habitat, which simulates missions on a mountain in Hawaii roughly 8,200 feet above sea level, in a relatively isolated location.[20] The crew members, who are largely amateurs or scientists and not astronauts, simulate communications with Mars by having a 20-minute time delay with their ground control — a practice also adopted by the Mars Society's Mars Desert Research Station in rural Utah, and sometimes by the NASA Extreme Environment Mission Operations Aquarius facility off the coast of Key West, Florida, where simulations are conducted underwater.

A 2021 study on HI-SEAS[21] examined how crews may self-organize during simulated missions to Mars.

Whereas mealtimes and daily call-ins on the ISS are designed to keep the crew on approximately the same schedule from day to day, HI-SEAS discovered a certain amount of schedule drift when the crews were left on their own. "The longer the missions went, schedules became more unpredictable over time," the paper noted. "Crew members stayed up later into the night, sought more privacy and time away from other crew members, and more distinct sub-groups formed for exercise and other leisure activities that previously were a full-group activity."

On Mars, such a situation may become even more pronounced given the difference in its day length from that of Earth. A "sol" or Martian day is 24 hours and 37 minutes, requiring humans to constantly shift their biological clocks to stay in tune with the sunrise and sunset. Typically, NASA asks its rover controllers on Earth to shift to "Mars time" for the critical first three months of a mission to get the most scientific data early on, in case problems develop later; but these controllers have reported it gets very challenging to keep their body happy with constantly adapting their earthly time zone by nearly 40 minutes a day.[22] We are still trying to figure out how to accommodate human physiology in the environment of a different planet, and will need to get that under better control before asking crews, or even earthbound controllers, to adapt their schedules to a new diurnal rhythm.

Being in space is its own reward, but there are challenges to consider when managing personal energy and managing time. NASA and its partners have made a lot of progress in the last 25 years, learning what works for the ground and what works for the crews. Future crews on the Moon or on Mars will likely have more autonomy, although how that will be achieved and accommodated is still being worked out through scientific studies and ongoing discussions at NASA. For now, however, the ISS serves as a well-oiled example of how humans can work with limited contact with Earth, as the astronauts can go for hours at a time managing their own schedules — a big step forward from the days of Skylab 4.

"Think of this as one big experimental vehicle — which it really is because it is the first of its type and one-of-a-kind," Lu wrote of the ISS.[23] "Our day-to-day operations, including repairs and maintenance, are giving us experience that will hopefully help us design and operate long-duration missions to asteroids and to Mars. Sometimes the lessons we learn are how to do things, sometimes the lessons are how not to do things. But we, as well as the engineers, managers, and scientists, are learning things that can help us leave low Earth orbit and explore."

A successful space mission often comes down to prudent management of time and energy, which is a kind of "space-time continuum." To be sure, time management for astronauts is not quite as physics-based as Einstein's theory, but there are valuable lessons to be learned from physics as well as simulation. Physics has shown us that time is relative and that it can depend on the observer. Good mission management allows a crew to feel "time rich" under many circumstances, and to give "more time" under emergency circumstances. As always, efficient use of time on mission will always depend on good training — especially when working with long-duration missions to the International Space Station and even more distant destinations.

CHAPTER 6

Are We There Yet?

Mars is the fourth planet from the Sun — it will likely be the third destination for humans in their quest to explore space. And it will be the most ambitious one so far.

Travelling to the 250-mile orbital altitude of the International Space Station is not particularly onerous. Rendezvous and docking generally occur the day after liftoff with the spacecraft travelling 25 times the speed of sound. At similar speeds it takes around three days to travel the 250,000 miles to the Moon — a longer trip than going to the space station but with considerably more possibilities to explore, greater opportunities for research after arriving and the bonus of being able to go outside for a walk — albeit in a spacesuit! But going to Mars is another story. At its farthest, Mars is 250 *million* miles from the Earth. This is not a trivial journey of a few weeks, it's a major commitment that will last a couple of years.

The good news is that Earth, the Moon and Mars are in elliptical orbits not circular orbits. Astronomers have suggested that elliptical orbits are common, the result of initial conditions in the formation of planets that make them more likely for the orbital characteristics.[1] If the orbit of the Moon around Earth were circular, the distance from Earth to the Moon would always be the same. In fact, the orbit of the Moon is a nearly circular ellipse, resulting in a slight difference in distance from the space traveller's perspective, depending on when you take off. However, the elliptical orbits of Earth and Mars around the Sun mean that at one point in their circuit each planet will be closer to the Sun than at any other time, and closer to each other. Coincidentally providing an important opportunity for a shorter voyage.

At its closest point to Earth, the Moon is 225,623 miles away, while the minimum distance to Mars is a mere 33.9 million miles.[2] For the Moon, the change in distance will not significantly affect the time to get there; but for Mars, waiting until the planets are closer together means roughly 216 million fewer miles to travel! That would shave months off the travel time — reduced deconditioning, less boredom and fewer opportunities for someone on the crew to ask, "Are we there yet?"

NASA and other countries' space agencies have used elliptical orbits to their advantage when sending rovers to Mars, and the same will be true with human missions. Because Earth and Mars orbit the Sun at different speeds, the launch of a Mars-bound spacecraft can be timed to minimize the distance travelled as it chases Mars to get there. Fortunately, the Earth is travelling around the Sun at 68,400 mph while Mars is slower, at 53,700 mph, making it easier to catch. Timing launches from Earth so that a spacecraft completes its long journey efficiently is critical. With the Earth and Mars orbiting the Sun at different speeds this occurs once every 26 months.[3]

Fundamentally, there are two different ways to get to Mars. One opportunity is something NASA calls the long-stay or conjunction class missions. These missions would launch from Earth when the distance between the two planets is the shortest. The astronauts would land and work on Mars until the planets are aligned again for an optimal return. This takes a little less than one Mars year, which is equivalent to two Earth years as Mars is farther from the Sun; of that time, 560 days would be spent on Mars. These missions are travel-efficient, around 198 days each way, and maximize the amount of surface time, but they require astronauts to be away from Earth for up to three years and some feel that might be riskier.

The other way to get to Mars is called an opposition class mission, launching from Earth when Mars is opposite the Earth in its orbit around the Sun. This option aims to take advantage of higher-energy propulsion systems to shorten the trip to Mars to roughly 180 days. Astronauts would spend between 30 and 90 days on Mars, leaving before the return window closes.[4] These missions might reduce the round trip to a little over a year and a half — the majority of which would be spent on the 342-day voyage travelling home. NASA is considering this plan for the first human mission to Mars, but it's a long way to go to stay for a month! Either way getting there and back will be a challenge. Essentially the decision will be based on reducing risk and optimizing benefit.

For some, there's nothing worse than a long, boring car ride. Just the thought conjures up childhood memories of staring at the ceiling of the vehicle trying to cover your face to stop breathing the smoke from your parents' cigarettes, or the more modern experience of watching a seemingly endless stream of DVDs while waiting for the next pit stop. The trade-off was the destination was so amazing it made getting there worthwhile. But getting there can be monotonous.

Near the end of the Apollo missions to the Moon, the astronauts commented on the tedium of getting there. During the last Moon mission, Apollo 17 commander Gene Cernan said of his voyage, "A funny thing happened on the way to the Moon: not much. Should have brought some crossword puzzles."[5] Perhaps "to boldly go despite the boredom" will be the biggest challenge for astronauts going to Mars.

It is easy to keep busy on the International Space Station. Despite the isolation, friends, family and the familiar activities we enjoy on Earth are only a VOIP (voice over internet protocol) call away. It is easy to stay in touch with, and sometimes participate in, current events; one NASA astronaut, Suni Williams, ran the Boston Marathon on the ISS treadmill during the actual event on Earth. Wearing bib number 14000, she finished the marathon with an official time of 4:23:10. While her terrestrial competitors had to endure the 48 degrees Fahrenheit temperatures with some rain, mist and wind gusts of 28 mph, the station weather was a balmy 78 degrees Fahrenheit with no wind or rain and 50 percent humidity.[6] Williams deserves special credit for total distance covered — roughly 78,950 miles while she, the treadmill and the ISS orbited the Earth at 17,500 miles per hour.

But the spacecraft required to get to Mars will not be as large as the ISS. Nor will they have all the amenities of a stationary lunar habitat that has the added fun-factor of going outside for a stroll or a ride on the lunar rover. The spacecraft will be small, there will likely not be much to do on the long journey, and the best option might be to sleep. But even trying to sleep can be a challenge as the astronauts still need to make time to eat and exercise.

Perhaps there's another option: hibernation. If astronauts could be like bears who hibernate over the winter, that might be long enough for them to sleep their way to Mars.

Bears can sleep more than 100 days without eating, drinking or passing waste![7] They survive cold winters by falling into a state called torpor, during which their heart rate, breathing rate and body temperature decrease. Bears are not the only organisms that can survive for long periods of time without food or water; the list of hibernators is quite long and includes bats, ground squirrels, mouse lemurs, snakes, leeches and marsupials.[8] It seems, however, that there is a difference between the torporists and the hibernators: true hibernators tend to be smaller creatures that can drastically reduce their heart rate and body temperature. Torporists, such as bears, are much larger and while their metabolism drops by 75 percent, their body temperature drops only a few degrees.[9]

Space programs have been interested in hibernation, torpor and suspended animation for years, but research programs have yet to find a practical solution for humans to sleep their way to Mars. There are many case reports of humans surviving after drowning in cold water, and modern clinicians routinely use body cooling to slow metabolic rates during cardiac surgery or use medically induced comas to help the critically ill survive. But those events are relatively short-lived and require a host of support technology, and for now they are not a practical solution for use in space. That said, the ISS partner space agencies are pretty good at turning science fiction into science fact, so perhaps the space pods portrayed in the movies *2001: A Space Odyssey*, *Alien* and *Passengers* will one day become routinely used for interplanetary travel.

One of the primary benefits associated with hibernation is the reduction in metabolic rate and need for food consumption. Heiko Jansen, an expert in integrated physiology who studies hibernation in bears at Washington State University, said in an interview with *WIRED* magazine, "The primary benefit there is that it allows us to be able to use a spacecraft that doesn't require carrying all the food that's necessary to bring someone to Mars, the passengers on the spacecraft could actually enter that state of hibernation and so be transported for who knows how long, maybe a couple of months even without having to eat anything." With a lowered metabolic rate during hibernation, "bears don't suffer any cardiovascular problems, they have no loss of bone density, they don't lose any muscle mass . . . So, there's a lot of interest in trying to understand . . . [those changes in] people that are bedridden or in space travel, [where] the loss of muscle mass is considerable." Apparently bears do not seem to experience that, even though they spend about 98 percent of the day laying around during hibernation.

John Bradford, president and chief operating officer of SpaceWorks Enterprises in Atlanta, has speculated that hibernation may be possible in humans, with a few caveats. Bradford, imagining a relatively short trip to Mars of six to nine months, proposed lowering the body temperature of astronauts by 9 degrees Fahrenheit to induce a "hypothermic stasis" that would decrease their metabolic rates by 50 to 70 percent. "That reduces the need for consumables in both nutrition and hydration, [and] oxygen demand," he said during a talk at the 2016 NASA Innovative Advanced Concepts (NIAC) conference. "That translates to mass, and mass is a critical item trying to support these Mars missions," he continued, adding that there might also be psychological benefits as the sleeping astronauts would not need to keep themselves entertained for long periods of cruise time in space.[10]

Bradford and his team have received multiple rounds of funding through the NIAC program, which examines far-range space exploration options to try to develop breakthrough technologies.[11] Their approach is to use the "therapeutic hypothermia" already available in hospitals to help people recover from traumatic injuries, slowing down the metabolism by keeping the body at near-freezing temperatures for hours at a time. Part of Bradford's work was to "initiate validation studies with leading medical researchers to understand the effects of prolonged hypothermia," with an eye to whether this research would help with future human space voyages.[12]

As Bradford suggests, perhaps one of the greatest advantages with sleeping during the travel phase of the mission is avoiding the day-to-day monotony of a seven-month voyage to Mars. While ISS astronauts can spend hours looking at Earth from their vantage point in space, the view from a spacecraft travelling to Mars will be much less compelling as Earth gets farther and farther away in the distance. Technology might provide a solution, either using personal immersive virtual reality devices or high-resolution TV screens that project images from Earth. (Although one can imagine the potential for heated discussions among the astronauts: "I want the beach in

Hawaii! No, we did that yesterday, I want the virtual walk around New York City . . .")

Living in space for long periods of time has revealed the behavioral challenges of having a small group of diverse individuals living together in a confined area. In the competency-based world of human spaceflight, the psychological aspects of the crew and how they work together are arguably more important than their technical competencies. Success depends on team dynamics, emotional stability, a sense of humor, resilience, optimism, tolerance of isolation and an ability to work together in small groups. Pete Roma, a research psychologist who has worked on NASA projects, commented in a 2016 interview, "A mission to Mars is much more analogous to an exploration mission at sea or in Antarctica than to an ISS flight. These people aren't only your co-workers, they're also your roommates. . . . There's always going to be some kind of conflicts."[13]

NASA funded the HI-SEAS program, presented in the previous chapter, to simulate a long-duration planetary surface mission, and the Institute for Biomedical Problems (IBMP) in Moscow has conducted the Mars-500 simulations — both agencies researching how to help future Mars-bound crews manage not only how they spend their extended time in space, but also the sleep disturbances, lack of privacy and sensory stimulation, monotony, discomfort, and potentially life-threatening situations that might occur.[14]

The longest Mars simulation, a 520-day study, took place in an experiment facility located at the IBMP site in Moscow from June 3, 2010, until November 4, 2011. It was the last of three separate simulations and had a six-member crew made up of Italian engineer Diego Urbina, French engineer Romain Charles, Russian physiologist Alexandr Smoleevski, Russian surgeon Sukhrob Kamolov, Russian engineer Alexey Sitev and Chinese astronaut trainer Wang Yue. The crew lived and worked in five different modules, with the habitat, utility and medical modules simulating the main spacecraft and an interconnected fourth module simulating a Martian lander. A fifth module connected to the lander was used to conduct spacewalks on a simulated Martian surface.

By all accounts the mission was a success. In the ceremony after the crew emerged from the hatch, Romain Charles said, "One year and a half ago, I was selected by the European Space Agency to be part of the Mars-500 crew. Today, after a motionless trip of 520 days, I'm proud to prove, with my international crewmates, that a human journey to the Red Planet is feasible. We have all acquired a lot of valuable experience that will help in designing and planning future missions to Mars."[15] It was clear that humor must have been part of making the mission a success when colleague Diego Urbina mentioned on Twitter the day before hatch opening, "The longest night in the world is about to finish . . . 'We come in peace,' I always wanted to say that."

Urbina spoke of the things he missed: "Simple things such as the blue sky, such as . . . going dancing in the evening. I love that. And here I am not able to do it. My family, I miss them a lot . . . I miss a lot the randomness of the world." According to one of the researchers, the crew worked together to find ways to break the boredom. "One thing that we did not foresee, but that they were using to break the monotony was creativity. So . . . for Halloween, for example, they dressed themselves up . . . for Christmas, they came up with their own self-made nativity scene, and they also celebrated the Chinese New Year, using Chinese folklore things."[16]

After the year and a half mission the crew felt proud to have achieved the longest ever simulation of space flight, so that "humankind can one day greet a new dawn on the surface of a distant but reachable planet."[17] However, not all simulations are as successful, and the psychological stresses of confinement can be significant. Ten years earlier, a similar simulation of a long-duration mission with international crew members had ended in conflict. There had been a fistfight and a separate incident that resulted in an allegation of sexual harassment following unwanted kissing during a New Year's celebration.[18]

Despite the overall harmony among the crew in the 520-day IBMP study, the researchers made important observations that could affect long-duration missions: "The majority of [the] crewmembers also experienced one or more disturbances of sleep quality, vigilance deficits, or altered sleep-wake periodicity and timing."[19] The many findings from the Mars-500 studies suggested the need for further research, which NASA sought to conduct in the HI-SEAS project.

The HI-SEAS habitat is located on the isolated slopes of the Mauna Loa volcano on the island of Hawaii. The area is desolate. The features of the 8,200-foot-high terrain look almost identical to images of the surface of Mars. From 2013 to 2018, NASA completed six missions there; the longest,

HI-SEAS IV, lasted a year. All the crews went through a rigorous selection process like that used to select astronauts, and the six-member teams lived in a solar-powered, dome-shaped habitat that had composting toilets, freeze-dried food similar to that used in space, limited medical supplies and a simulated 20-minute delay on all communication with the outside world, the same as the delay between Earth and Mars.

In an interview with *National Geographic*, HI-SEAS principal investigator Kim Binsted commented, "This is about crew cohesion and performance, so how do we keep a crew cohesive? How do we select a crew and train a crew so that they can be resilient? What we've found is that there's no magic bullet to prevent conflict, it's how you deal with it and how you respond to it. Not just as individuals, but as a group."[20] Commander Carmel Johnston described the crew's challenge: "But how you deal with that in a dome or in a confined space is much different than if you can just walk away. We want to learn everything that can go wrong before it goes wrong in space and prevent it from happening."

The project was tough for all the crews, exacerbated for some by experiencing deaths in their families, while others missed weddings and births. Despite the challenges they had a few successes, including the use of virtual reality to help deal with the isolation. Virtual reality is thought to reduce feelings of loneliness and has been studied with seniors to determine if it can elicit feelings of connection, support and well-being.[21] The results have confirmed its value and it will undoubtedly play a role in future long-duration spaceflights.

The outcome of the sixth HI-SEAS mission was not as positive. The mission started February 15, 2018, and the first few days were cloudy, making it hard for the solar panels to charge the batteries that powered the station, though a propane-powered generator was available as a backup. By February 19, the batteries were dead and the crew donned their simulated spacesuits to start the generator.

The start-up was successful, but one of the crew was inadvertently electrocuted when they flipped a switch on a circuit breaker inside the habitat. After considerable discussion about breaking the isolation protocol and unsuccessful attempts to immediately reach the mission support team, the crew called 911 and the injured crew member was taken to hospital for medical treatment. For a number of reasons, the mission was terminated after the incident and the entire team took the opportunity to learn about the risks that can threaten a mission to Mars.

"We've learned all the ways that you can kill yourself on Mars, and we've learned to prevent those things," said Bill Wiecking, the HI-SEAS technical support lead, in a 2018 interview with *The Atlantic*. "So it's been very, very valuable, because it's way better to do it here, where you can drive up and go, 'Oh gosh, a water valve opened up and now you don't have any water.' Instead of on Mars, where it's like, 'You don't have any water, you guys are gonna die in a couple of days.'"[22]

Drawing on the shared lingo developed by Antarctic explorers, there is no question the A-factor (the Antarctic factor, where unexpected extra difficulties arise during an expedition) will become the M-factor — the Mars factor — on future voyages, and teamwork and psychological resilience will be more critical than ever to mission success. In the frozen world of Antarctica, with its demanding mission objectives, harsh conditions and isolation, when modern-day explorers are having "a bit of a boot today" (are down in the dumps) or are "epischeded" (exhausted, finished, dead beat, done for, knackered), having some "freshies" (fresh fruit and vegetables) can make a big difference to outlook and morale. Without a doubt these experiences and the lexicon of extreme exploration would resonate with long-duration crews on the ISS and those participating in missions beyond Earth orbit.

The most fundamental scientific question that remains unanswered is whether life exists elsewhere in our solar system, the Milky Way galaxy or the universe. The huge distances belittle the current propulsion technology of modern spacecraft, so sending humans means the search for an answer will start close to home, by galactic measures. There is no question that missions to Mars will bring a host of technical, medical and behavioral challenges, possibly even including the death of a crewmember; but many would argue that the scientific benefits and new discoveries outweigh the risks as humans reach out to explore the fourth rock from the Sun.

CHAPTER 7

Working Like an Insect

The first step is tentative. One foot is lightly placed upon the ground, and once it is clear that the surface is safe, the next foot follows. Then there is typically a pause, during which every sense is used trying to determine any danger that might exist. None detected, the cockroach emerges from a gap around the sink drainpipe to search for food.

There are between 3,500 and 4,000 different types of cockroaches throughout the world, and depending upon where you live there can be over 30,000 existing in one building. They seem to be perfectly adapted to survive in their environment, with a 300-million-year history of living on Earth. They are members of a group, or phylum, known as the arthropods. This is the largest phylum in the animal kingdom and includes cockroaches, other insects, crabs, spiders, centipedes and millipedes. They don't have backbones; one of their differentiating features is an exoskeleton that has a segmented body and jointed appendages, somewhat like a medieval knight — or an astronaut!

Exoskeletons have several advantages. Probably the most important is their ability to provide protection. The challenge with exoskeletons is moving around, especially at the speed of the cockroach, which has been clocked at over 5 km/h (3 mph).[1] That may sound slow, until you realize that if the six-legged cockroach were the size of a human, that would be a speed in excess of 100 mph![2] Cockroaches have perfected the art of movement thanks to the muscles in their limbs being directly attached to the exoskeleton, rather than to bones as in other animals. This allows them to transition from walking on just three legs, using two on one side and one on the other side, to alternating between three legs on one side and all three legs on the other side. No wonder they're fast when they have six legs working for them.

On the other hand, a study has shown that a medieval knight walking in armor at a fast pace could reach only 1.7 meters per second, or 3.8 mph. To do so for any distance proved to be a challenge due to the weight of the armor, the difficulty breathing and the challenge of heat building up inside the suit.[3] Speed is an important element of mobility, and while it would be hard to have an actual race, is an astronaut in a modern spacesuit faster than a medieval knight? Based on the results of NASA research, it seems like astronauts in their bulky spacesuits could beat a knight in armor — at least in a simulated lunar 10K.

During the later Apollo missions, astronauts used the lunar rover to search for geological specimens. There are spectacular photos of the rover perched on the edge of a huge crater and videos showing the astronauts throwing up rooster tails of lunar dust as they tested its top speed. In getting ready to go back to the Moon, NASA researchers are asking, "What would happen if the rover broke down? Could the astronauts walk back to their lunar spacecraft?"

The team at Johnson Space Center did a study to determine the feasibility of suited astronauts performing a 10K walkback in simulated Earth, lunar and Martian gravity. Designed to determine the feasibility of walking long distances wearing a spacesuit, the test also collected human performance data to understand the exercise level and biomechanics while test subjects walked in the suit at different speeds.[4]

For the subjects, this wasn't your average walk in the park. They were given 32 ounces of water to drink from a specially designed bag inside the spacesuit. There were no snacks, and during the roughly 90 minutes it took, they used from 931 to 1,068 calories (most of us would burn less than 500 calories walking the same distance). To recreate the gravity of the Moon and Mars, a special device called a POGO was used to partially support the weight of the subject wearing the spacesuit. Perhaps getting the lunar 10K T-shirt was a sufficient motivator, as two of the astronaut test subjects achieved a top speed of 5.5 mph running with gazelle-like strides in simulated lunar gravity, similar to the speed of many joggers on Earth.

The study showed that it is feasible to walk back from a rover that is 10K away from a lunar habitat, reducing the need to launch a backup rover to go and get stranded astronauts. Astronauts are typically up for any task; however, their debrief comments illustrate some of the difficulties of walking and working in a spacesuit. They would have liked more water to drink, an energy bar or gel for a snack, improved suit biomechanics and reduced friction points. It became clear that humans are not ideally designed to work inside a suit. For us, having an exoskeleton is a challenge.

The first spacesuits were designed to protect the astronauts in case the pressurization system of their spacecraft failed. For the original Mercury astronauts, the cabin pressure of the spacecraft was maintained at 5.0 psi, equivalent to an altitude of close to 25,000 feet on Earth, and the oxygen was critical for survival. The suit was derived from the high-altitude pressure suits worn by pilots and weighed 22 pounds. It had an internal pressure of 4.6 psi[5] while its user breathed 100 percent oxygen and relied on the life support systems of the capsule, thus requiring no heavy, external oxygen supply. It was quickly determined that the pressure inside the suit had to allow the astronaut to move their arms, legs and most importantly their fingers.

Fortunately, the Mercury astronauts never had to rely on their spacesuit to survive a failure of the capsule pressurization system. But to achieve the objective of landing humans on the Moon, NASA had to develop and test a spacesuit that would allow astronauts to work outside their spacecraft in the vacuum of space. That task fell to Ed White, the first American astronaut to perform a spacewalk.

His spacewalk took place in June 1965, and while he was the first American to venture into the vacuum of space, he was not the first person to do so. Close to three months earlier, Russian cosmonaut Alexei Leonov performed the historic first spacewalk in a mission later referred to as a "nightmare."[6]

Leonov exited the Voskhod 2 space capsule airlock to float freely in space for 10 minutes before coming back inside. His initial reaction was one of freedom and exhilaration, and said he felt "like a seagull with its wings outstretched, soaring high above the Earth."[7] When fellow cosmonaut Pavel "Pasha" Belyayev asked him to come back inside the capsule, it reminded him of his mother calling him to come back inside when he was outside playing with his friends.

At that moment, Leonov learned the challenge of dealing with a suit that is more like an exoskeleton than an item of clothing. With the spacecraft travelling 25 times the speed of sound, darkness was rapidly approaching and he realized "how deformed my stiff spacesuit had become, owing to the lack of atmospheric pressure. My feet had pulled away from my boots and my fingers from the gloves attached to my sleeves, making it impossible to re-enter the airlock feetfirst."[8] In the parlance of modern spacewalkers, "if something goes wrong when you're outside, you may have the rest of your life to solve the problem."[9]

It was critical that Leonov find another way to get back inside before he ran out of time. If feetfirst in a puffed-up suit wasn't going to work, he would have to find another solution. In a maneuver somewhat reminiscent of trying to put a cork completely back into an open bottle of wine, he realized the tight fit would require him to bleed off some of the oxygen from his spacesuit. He would either lose consciousness from oxygen starvation or he would succeed. Either way the clock was ticking as he had only 40 minutes of oxygen remaining in his life support system. The exertion of his effort caused the temperature to rise dangerously within his suit and his visor became fogged, adding to the difficulty of re-entering the spacecraft, but he eventually succeeded despite the challenges. His experiences helped improve future Russian suit and airlock design, but unfortunately were not known to NASA; Ed White experienced similar problems with visor fogging during his time outside his Gemini capsule.

The Voskhod 2 spacewalk was only the beginning of Leonov's and Belyayev's problems, however. As they readied themselves for re-entry into Earth's atmosphere, they realized that their automatic guidance system was not working properly, and they did their best to orient the spacecraft manually. Unfortunately, they landed long — more than 900 miles west of their predicted landing site — and spent a winter night in deepest Siberia. The forest where they landed was inhabited by bears and wolves. With a pistol, a lot of ammunition and a spacecraft reconfigured for survival, they made it through the night.

Gene Cernan was the third person to walk in space, during the Gemini 9A mission. With only 20 minutes' experience from Ed White's spacewalk to build on, the tasks planned for Cernan's proved to be particularly ambitious. After exiting the Gemini capsule, he was to move to the back of the adapter section and strap on an astronaut maneuvering unit, disconnect his oxygen umbilical from the spacecraft and fly around the Gemini while remaining attached with a tether. He experienced the same problems Leonov had encountered with limited movement in the spacesuit: lack of handholds, overexertion, overheating, and fogging of his visor. Using the tip of his nose to clear a small area so he could see, he was able to get back into the capsule, struggling with the suit the entire time.[10]

Solving the spacewalking challenges of the Gemini astronauts would be critical for NASA crews to work successfully on the surface of the Moon. Despite efforts to resolve the issues experienced by Cernan and White, Mike Collins and Dick Gordon had similar challenges during their Gemini spacewalks. This resulted in Dr. Robert Gilruth, director of Johnson Space Center (then the Manned Spacecraft Center), telling the astronaut office to use "both zero-G trajectories in the KC-135 and underwater simulations"[11] to help prepare the astronauts working in space. The astronauts had been evaluating the use of underwater training to simulate the microgravity of space, and the KC-135, a zero-G aircraft also known as the "vomit comet," had been used for training in 20-second blocks of microgravity during parabolic flight.

With the Gemini program rapidly coming to a close, Buzz Aldrin and Jim Lovell used the Gemini 12 mission to test new approaches to spacewalks. Aldrin worked hard during underwater training to develop new processes, including greater use of handrails, use of a waist tether to help hold the astronaut at the work site, and foot restraints. When he combined these new techniques with a slow, deliberate approach to moving alternated with moments of rest, he finished the simulated two-hour spacewalk without fatigue. Looking ahead to Apollo, Gemini program manager Walt Williams said, "It is time to go on. We will be able to go on with confidence because there was this program, and the program was called Gemini."[12]

The Apollo astronauts spent a total of 80 hours and 32 minutes exploring the lunar surface during the six missions that landed on the Moon.[13] They performed a number of scientific tasks, collected geological specimens and evaluated the driving characteristics of a lunar rover. The missions were incredibly successful, historic achievements that set a high standard for future human spaceflights. They were also opportunities to learn more about the complexities of working in a spacesuit in lunar gravity, which is one-sixth that of Earth. The astronauts had to deal with ever-present dust contaminating their suits, the tendency to fall forward (suggesting that the center of gravity of the suit was too high and too far forward) and the fatigue of moving in the suit while they worked hard to perform several different tasks.

Despite these challenges, the Apollo astronauts were remarkably successful in accomplishing their objectives within the scheduled timeline. The range of tools they used provided a breadth of experience that would be important in preparing for future spacewalks on the Skylab space station. These lessons were also incorporated into the design of a new spacesuit called the extravehicular mobility unit for the space shuttle and space station programs.

Modern spacesuit design has improved mobility significantly from the suits worn by the Gemini and Apollo astronauts. The name itself — extravehicular mobility unit — emphasizes the importance of being able to move around freely and efficiently while working with tools and allowing sufficient dexterity to perform different tasks. It is a tall order for the suit designers and engineers, since the extravehicular demands placed on the suit mean that, in essence, it has had to evolve from a balloon-like exoskeleton to a flexible, agile second skin. Today's modern spacesuit has been tested with 14 different joint and body movements to enhance suit performance and increase the range of motion and useful work envelope of the suit.[14] It is essentially a one-person spacecraft that has two major subsystems, the life support subsystem and the suit itself, otherwise known as the space suit assembly.

Getting into the suit is a lengthy process that starts with donning the maximum absorbency garment (MAG), otherwise known as the diaper. Spacewalks last between five and eight hours depending upon the tasks, and there is no rest stop when tethered to a craft that is travelling five miles a second. Despite the ever-present availability of the MAG,[15] few astronauts end up using it. It's easier, and more pleasant, to visit the restroom before putting the suit on. Also, working in the sealed, 100 percent oxygen environment and moving repeatedly against the 4.3 psi pressure of the suit can be a demanding and dehydrating experience. Most astronauts drink the entire 32 ounces of water available in their in-suit drink bag and still find themselves so dehydrated that they don't feel the need to use the diaper.

Next, astronauts put on long underwear, followed by a white mesh item called the liquid cooling and ventilation garment. It incorporates plastic tubing, reminiscent of that used in small aquaria, that circulates cold water to prevent temperature building up inside the suit during demanding tasks. Based on individual preference, astronauts can choose to add extra padding to the garment's shoulders, elbows and knees.

The spacesuit has a hard upper torso (HUT) made of fiberglass that holds the portable life support system, a backpack look-alike where oxygen containers, carbon dioxide scrubbers, a battery and the cooling system are located. The five-layered flexible arms and legs attach to the HUT, and the helmet is attached once the astronaut is in the suit.

Astronauts like to joke that getting into the suit is like a turtle trying to crawl back into its shell. The legs, or lower torso assembly, are put on first. In space, it's a little easier when floating around to slide into the legs of the suit and pull the waist ring up above one's hips. The next part is considerably more difficult. Getting into the HUT requires flexibility and patience, and is not for the claustrophobic! Rivalling the best yogis, the most commonly used approach begins with both arms above the head reaching into the HUT to find the shoulder openings. With steady gentle pressure, the arms continue into the sleeves and the head slowly moves upward inside the rigid torso of the suit. If everything goes according to plan, the arms can be used to pull the upper body into the suit with one smooth movement as they are brought down to the side of the suit. The head then emerges from the opening at the top of the HUT, and it's time to attach the cooling garment to the water supply and close the waist ring.

After squeezing into the HUT, the remainder of the donning process is straightforward: putting on the helmet and the two gloves. It is nice to have someone help with the helmet and gloves, but it can be done solo if necessary. With the helmet on, the airlock service and cooling umbilical are attached to the front of the suit to provide oxygen, power and cooling for the next phase of preparation — pre-breathe.

Astronauts have a 100 percent oxygen pre-breathe period before the spacewalk to reduce the risk of getting decompression sickness, also known as the bends, a problem normally found in scuba divers. It occurs when the diver returns to the surface after a dive and the surrounding pressure is reduced. The nitrogen in the air they breathe during the dive gets dissolved in their tissues, and the increased pressure at depth increases the amount of dissolved nitrogen. If they return to the surface too quickly, the nitrogen is rapidly released from the tissues, causing bubbles to form in the bloodstream and in the tissues themselves, which can be fatal. Similarly, when astronauts transition from the pressure inside the space station, which is the same as it is on the surface of Earth, to the lower pressure of the spacesuit when the airlock is depressurized, and then to the vacuum of space, they run the risk of getting the bends. Anytime the human body goes from higher to lower pressure quickly, nitrogen bubbles can form that can either mechanically damage the tissue or block off circulation to the tissue. To prevent nitrogen buildup in the bloodstream and tissues, astronauts perform a pre-breathe before going into the airlock to reduce the amount of nitrogen in their body, in one of several ways.

If there is enough time, at least four hours, it is possible to breathe 100 percent oxygen in the suit before the spacewalk. This procedure can work for planned spacewalks but is problematic in emergencies or other situations when astronauts may need to get outside more quickly (including future lunar spacewalks where going outside will be routine). Another option is to reduce the pressure of the entire space station, spacecraft or habitat for a period of time before the in-suit pre-breathe. This can reduce the pre-breathe time to 40 minutes if the pressure has been lowered for more than 24 hours. Yet another option is to combine spending time at a lower pressure with exercise, or simply exercising while breathing 100 percent oxygen, to increase the rate at which nitrogen is removed from the body. Future lunar habitats may be kept at a consistently lower pressure than that of the current space station to improve the efficiency of getting ready for a spacewalk.

After the pre-breathe, the astronauts are put in the five-by-nine-foot crew lock.[16] Once the hatch is closed it is slowly depressurized until it is equivalent to the vacuum of space, so they can safely open the outer hatch to start their spacewalk.

There have been 241 spacewalks performed on the space station between 1998 and 2021.[17] Fortunately, no one has been hurt, but there have been emergencies. In 2007 NASA astronaut Rick Mastracchio was outside the space station with colleague Clay Anderson during the STS-118 shuttle mission to build and service the station when he noticed a hole in the outer layer of one of his spacesuit's gloves.[18] Later inspection of the glove suggested it was cut by a sharp edge created by orbital debris hitting the space station. Orbital debris, or space junk, is the result of man-made objects that no longer have a purpose, such as fragments of degraded spacecraft or retired satellites. It creates a significant problem that can affect the space station, satellites and, potentially, spacewalking astronauts. The risk of spacewalkers being hit by a hypersonic particle varies depending upon where they are working on the space station, with an overall risk of 1 in 7,000 for any size leak being caused and 1 in 35,000 for catastrophic leaks.[19]

If the risks of decompression sickness, being hit by orbital debris or suffering from a pulled muscle or other injury don't underscore the potential dangers of spacewalking, then the story of Italian astronaut Luca Parmitano certainly does. Thirty to 45 minutes into his July 16, 2013, spacewalk, the helmet of his spacesuit started to fill with water: "I felt some water on the back of my head. I realized it was cold water, it was not a normal feeling. I started going back to the airlock and the water kept coming in. It completely covered my eyes and my nose. It was really hard to see. I couldn't hear anything. It was really hard to communicate. I went back using just memory, basically going to the airlock until I found it."[20]

NASA referred to the incident as a "high visibility close call" and the subsequent incident review found the source of the water leak was a clogged filter. The suit was serviced in space and the problem did not recur. These incidents are opportunities for NASA to prepare for the upcoming lunar return missions, where the frequency and complexity of lunar spacewalks will build on the experiences of the Apollo, shuttle and space station astronauts. Future spacesuits might move away from the traditional rigid, pressurized suit, which makes it difficult to move, to a more flexible suit. Perhaps it's time to evolve from the exoskeleton model to a sophisticated second skin called the BioSuit.

Designed by Massachusetts Institute of Technology engineer and former NASA Deputy Administrator Dava Newman, the BioSuit uses a tight-fitting elastic fabric that can exert pressure equivalent to the traditional 4.3 psi spacesuit.[21] Not only would the skintight suit significantly enhance mobility, but a small puncture of the BioSuit could be easily repaired with a high-tech elastic wrap, unlike the potentially catastrophic leak that would occur in a pressurized suit. Newman feels it is time to "move beyond the balloon"; through ingenuity and research like hers, engineers hope to develop next-generation technologies that enable humans to work on the Moon and Mars.

Newman's team started by assessing movement within the current exoskeleton-style spacesuit. In a 2013 interview with TED Women, she said, "We have a wearable octopus [to sense motion] . . . so for the first time we're going to measure human motion within the suit, how the astronauts are moving within the suit, we can quantify that."[22] Evaluating the biomechanics 'and forces required to perform tasks while wearing the conventional suit helped the team envision a completely different concept for suit design. Rather than using the traditional pressurized spacesuit concept, Newman thought that a force equivalent to that of the pressurized suit could be applied directly to the skin by a tight, form-fitting suit: "It's not a gas-pressurized, 140-kilo system, [but] no I'm going to shrink-wrap you . . . We're going to put the pressure directly on your skin, and not have to provide 30 percent of an atmosphere [that] keeps you alive in the vacuum of space . . . and I want to give you maximum mobility."

She describes the suit design as "a wonderful aesthetic solution to give maximum mobility that requires minimum energy expenditure." Mobility will be critical to enable future Mars astronauts to search for signs of ancient life. In a 2019 podcast, Newman spoke with host Lex Fridman about exploring Mars: "Since we're going back to the Moon and [on to] Mars we need a planetary suit, we need a mobility suit, so that's where we've kind of flipped the design paradigm. I study astronauts, I study humans in motion, and if we can map that motion, I want to give you full flexibility . . . rather than a gas pressurized shrinking [of] the space craft around the person, can I design a spacesuit literally from the skin out? That's what we've come up with mechanical counter pressure. . . . and that can be an order of magnitude less in terms of the mass, and it should provide maximum mobility for Moon or Mars."[23]

Perhaps by 2029, the 60th anniversary of the Apollo 11 Moon landing, we will have humans on the lunar surface once again, continuing the scientific legacy of the Apollo missions while developing and testing new habitability systems, spacesuits, rovers and operational capability for future flights to Mars. More than 600 years after coat-of-plates armor was developed in the Middle Ages, humans will remain dependent on a protective suit to help them explore potentially hostile environments.

Unlike the knights of days long gone who sought protection from their foes, however, the hostility will come from the extreme, harsh conditions of space. Perhaps one day humans will be able to walk outside their extraterrestrial habitats the way we go outside on Earth, but for the imminent future, survivability will depend on the sophistication of the spacesuit design and the mobility it provides.

CHAPTER 8

Skin, Hooves and Nails

What do horses, guitarists and astronauts have in common? The question sounds more like the beginning of a joke told over a round of drinks than an important issue in space medicine. In fact, it's a question that NASA flight surgeons asked when trying to solve the problem of spacewalker's nails — otherwise known as fingernail delamination, associated with repetitive finger stress working in a spacesuit.

Fingernails are part of the keratin trinity: skin, hair and nails, all of which are important but taken for granted in our day-to-day lives.[1] Keratin is a protein that is also commonly found in the animal kingdom in hooves, claws and horns. Depending upon how the proteins are organized, keratin can form the thin, flexible barrier that protects the skin, the largest organ in the human body, or it can create luxurious hair and rigid nails.

From a biological perspective, humans have fingernails because we're primates.[2] Other mammals have claws that are used to grab, climb, dig and help with self-defense. The dexterity of human fingers is like that of other primates, whose fingertips make it easier to pick up smaller objects while the fingernails help protect them. Around 2.5 million years ago humans started to use stone tools, and over time the fingertips of humans became broader than that of other primates to help them perform more complex grasping tasks.[3]

Keratin is common to both fingernails and hooves, forming a layered structure that is either flexible or rigid depending upon the relationship between the keratin proteins and the layers of keratin. The outer part of a horse's hoof is called the wall. It is made up of layers of keratin, providing a hard outer covering that supports the weight of the horse and protects the more delicate structures of the underlying hoof.[4] Just like human fingernails, the hoof wall does not have nerves or blood vessels in it and is continually growing at a rate of about three-eighths of an inch per month. That's approximately three times faster than human fingernails, which typically grow at a rate of roughly one-eighth of an inch (3.5 mm) per month.[5]

It's amazing how much we can tell about a person's general health from their fingernails. Their rate of growth and overall appearance can be a clue to different illnesses, a phenomenon eloquently described by Dr. William Bean in a scientific paper entitled "A Discourse on Nail Growth and Unusual Fingernails," published in the early 1960s.[6] The paper covers unusual conditions such as onychogryphosis, which Bean describes as a "cheerful name given to the long, horn-like nail which may grow when the nails are not trimmed," and is illustrated by a photo in his paper showing the two big toes of a young woman who had stopped trimming her nails. The curved toenails look more like the horns of a ram than human toenails.

Bean conducted a personal observational study in which he spent 35 years monitoring his own nail growth, which he creatively described as a "slowly moving keratin kymograph that measures age on the inexorable abscissa of time."[7] Bean came to appreciate the various afflictions of fingernails and toenails, as well as the subtle clues they can provide to detect other diseases. Nails can crack, their layers can separate, they can become ingrown or infected. They can be beautifully painted or simply taken for granted when they are healthy. But they become a painful distraction if not properly cared for.

The medical term for split nails is either onychoschizia, if the split is horizontal following the edge of the nail, or onychorrhexis for lengthwise splits or ridging of the nail.[8] Split nails are relatively common, affecting up to 20 percent of the population, and can be caused by repetitive damage at work, exposure to various chemicals, frequent wetting and drying, or various medical problems.[9] And they are a common problem for both astronauts and guitarists.

Future spaceflights back to the Moon and ultimately to Mars will require astronauts to spend more time outside their habitat conducting scientific experiments than astronauts on the International Space Station currently spend performing maintenance spacewalks. As we saw in the previous chapter, the proposed planetary spacesuit design is an evolution of the current spacesuit and, although novel prototypes have been suggested, these suits will most likely be pressurized. Mobility is the greatest challenge working inside a spacesuit. When pressurized, the suit forces the limbs, gloves and fingers to become fully extended rather than remaining in their natural position.

In the case of the spacesuit gloves, for example, grasping any object requires sufficient strength — first to overcome the resistance of the extended pressurized fingers of the glove, and then to hold onto whatever object is being grasped. Success working in the suit depends on a good glove fit and the biomechanics of the astronaut's fingers moving inside the fingers of the glove.

The human hand has been described as "one of the most complex and beautiful pieces of natural engineering in the human body. It gives us a powerful grip but also allows us to manipulate small objects with great precision. This versatility sets us apart from every other creature on the planet."[10] The resting position of the hand is with the fingers partially flexed. Straightening, or extending, the fingers uses muscles on the top of the forearm. Flexing the fingers uses the opposite muscles on the lower portion of the forearm. Both actions require muscular contraction.

The resting position of the hand in a pressurized spacesuit is somewhat different. At rest, the pressure in the glove keeps the fingers fully extended; it takes more muscular effort than usual to flex the fingers. Every flexion requires force, and repeated force can lead to fatigue. It can also cause repetitive pressure on the fingertip and the fingernails. Due to the position of the fingertip in the glove as the finger flexes, contact with the glove pushes against the fingernail with a force that lifts it away from its nailbed. Similarly, the force is distributed along the edge of the multilayered fingernail, creating a stress that can cause horizontal separation of the nail.

This may all sound like much ado about a simple, minor ailment — sort of like complaining about a hangnail — but experts in space medicine would diagnose it as biomechanically induced spacewalker's onychoschizia!

And in space, nail problems can be more than just irritating: they can be painful, result in infection, and can easily get in the way of the astronauts accomplishing the important work they need to do. Kind of like a baseball pitcher with a blister — a pesky annoyance to most of us, but an occupational hazard that benches the major leaguer.

Protective comfort gloves are worn inside the spacesuit gloves to help reduce friction and pressure points, but they can also increase the moistness of the hand and fingers, which increases the chance of nail separation. Also, some individuals are more prone to brittle nails. The number and complexity of tasks during a spacewalk may contribute as well since the number of finger flexions over time increases the force applied to the fingernail. Filing nails and keeping them short can help, but most spacewalkers at some point in their training have experienced the problem.

Generally, the separation resolves as the nail grows out over a couple of weeks, but there is always the risk of developing a fungal nail infection between the separated layers that can be difficult to treat. The best solution is to prevent the separation, and the flight medicine team has undertaken a wide search for innovative preventive strategies.

Considering the biomechanics underlying horizontal nail separation, it is not surprising that fingerpicking guitar players have similar issues. Repetitive force to the edge of fingernails from guitar strings can have the same effect as the spacewalking glove, just as it can to the edges of horse hooves. The space medicine doctors wondered if guitar musicians or equine veterinarians, well-versed in treating cracks and separation of the layers of horse hooves, might have a solution that could help the astronauts.

Little did musician James Taylor realize that in sharing his strategies he might be helping spacewalkers: "Human nails are not strong enough to put up with the kind of repeated punishment that this fingerpicking style puts them through."[11] He "eventually went to a nail salon and had them put on a set of fake nails. . . . A lot of people do. But I find that it's such an important part of my performing kit . . . I need to be able to put nails on myself or to touch them up to keep them working." Guitarists are meticulous in caring for their nails and there are websites and numerous articles advising different aspects of their care.

Those with fragile nails are advised to "strengthen just the tips by painting on nail glue and dipping your fingertip into nail acrylic powder.

The powder sticks to the glue and becomes rough and semi-transparent. Sand this layer smooth. Apply a second layer of glue . . . when it is dry, sand and buff the edges."[12] However, the glue and powder reinforcements sometimes remove the outer layer of the fingernail itself with it when the protective coating falls off, thus creating a continuous cycle of using the glue and acrylic powder. But the consensus is that it makes for nice, strong nails for playing.

Many astronauts use nail hardener to strengthen their nails, and anecdotal observations suggest that Hard as Hoof nail strengthening cream works well. The benefit of using topical equine products for humans is gaining popularity; in fact, some are convinced the best nail cream for brittle nails is Healthy Hoof cream, meant for horses.[13] With hooves and fingernails both made from keratin, it is not a surprise that what works for a horse might help a human! If nothing else, it will make for an interesting conversation in the lunar habitat when one of the astronauts puts it on before a spacewalk.

Some feel that one of the benefits of space travel is that corns and calluses are no longer a bother in the absence of gravity. The characteristic hard, thick areas of skin can form anywhere on the body but occur most often in areas of frequent friction, irritation or pressure — typically the weight-bearing part of the foot, or on the heel or toes if shoes are particularly tight. Floating around in a space station sounds like a perfect solution. No shoes are needed unless you're running on the treadmill or cycling on the exercise bike. Otherwise, astronauts wear white socks. Sounds like blissful relief!

But calluses can form in space, though they are found on the top of the feet rather than the bottom or heel. While it is possible to float in a specific location in microgravity, for some tasks astronauts tuck their forefoot into restraining loops to stop themselves from floating away. Contact with the foot loop results in a point of friction on the top of the foot that over time may result in the formation of a callus.

Returning to Earth, astronauts find that the soles of their feet are smooth, while the tops are thick and callused like the soles were before the mission. This can be very uncomfortable when they start to wear shoes again, but over a couple of months the change in friction and weight-bearing shifts the areas of skin thickening back to the soles of their feet. Calluses can also build up on astronauts' hands when they are in space. It's easy for them to move about the space station by pulling themselves hand over hand, gliding gracefully between handholds, but using their hands so much for propulsion means that their skin thickens there instead of on their feet.

Changes to the skin are part of life in space. It is the largest organ in the body — adults have about 8 pounds (3.6 kilograms) and 22 square feet (2 square meters) of it.[14] It's a waterproof protective layer that shields the body, helps prevent infection and produces vitamin D. It is made up of three layers: the outer epidermis, the underlying dermis and the deepest layer, the hypodermis. The keratinocyte cells of the epidermis are continually renewing the top layer of cells, which flake off and are replaced by newer cells in a roughly five-week cycle.

That shed skin is sometimes attributed to creating up to 70 percent of household dust, but that number is likely exaggerated.[15] Most dead skin is washed down the drain when we bathe, but in space there's no shower and no gravity for the particles to fall to the floor — the dead skin simply floats in the air.[16] The NASA Skylab space station in the 1970s and the more recent Russian Mir space station had a showers but there isn't one on the ISS. The challenge of pulling the cylindrically shaped shower from floor to ceiling, spraying yourself with less than six pints of water and vacuuming the soapsuds from your body was not particularly time effective or convenient for the Skylab astronauts.[17] Today's ISS astronauts take sponge baths using liquid soap, water and rinseless shampoo.[18]

In the absence of a shower, astronaut Don Pettit has graphically described what happens when you take your socks off on the space station:

> You're carefully pulling the sock off . . . kind of teasing the sock off, and the reason you do this is to minimize the amount of particles coming out. But then you get to a point where the sock just pops off your toes, and this cloud, this explosion of skin particles, detritus, floats out and you're in this weightless environment and the particles have nowhere to go but out. And they're floating around the atmosphere, which is why you want to pull your socks off, or your shirt for that matter, in front of a filter inlet. So, all those particles are sucked against the filter and then sometime during the week you can come by with a vacuum cleaner, you can clean all those things up.[19]

Fellow astronaut Mike Massimino summed up the issue of floating clouds of dead skin most succinctly: "This sounds pretty disgusting."[20] Disgusting or not, it is simply part of day-to-day life on the space station.

And speaking of disgusting, there's a lot of room in 22 square feet for microorganisms to grow. We all have different bacteria and fungi that live on our skin. It is estimated that there are roughly 1,000 different species of bacteria that live there and over 80 different types of fungi.[21] This thriving community of microbes is called a microbiome — the collection of living things in a defined environment — just as it is in the gut. If one takes into account all of the microorganisms on the skin and inside the digestive tract, "human microbiota consists of 10–100 trillion symbiotic microbial cells harbored by each person."[22]

Everywhere humans go, they carry around their own little world of microorganisms which interacts with whatever new environment they are in. That includes visiting the space station. Researchers have studied the interaction of the human microbiome with the ISS environment. "We identified changes in the skin microbiome very early during the mission, and the change that we observed, independent of other changes in diversity, was a significant reduction in a type of bacteria that's called proteobacteria," commented Dr. Herman Lorenzi in a recent interview.[23] "More specifically, two subtypes of proteobacteria, called beta and gamma proteobacteria."

These bacteria are thought to help protect against developing hypersensitivity skin reactions, and perhaps their decrease is due to shedding of the bacteria with dead skin that occurs over the first few weeks an astronaut is in space. According to Lorenzi, "This type of proteobacteria is found in very high quantities in soil. Since the space station is a non-green environment that stays very clean, it is possible that the reduction in contact with Earth environmental bacteria reduces these microbes on the skin."[24]

The microbiome of the space station reflects that of its occupants, Lorenzi said: "The space station environmental microbiome looks very similar to human skin microbiome." The ever-present bacteria brought to the ISS by the astronauts themselves have resulted in a unique world of microbial tourists. "What was interesting in our study was that the composition of the station microbiome was changing all the time," Lorenzi said. "The station microbiome tended to resemble the composition of the skin microbiome of the astronauts that were living in space at that particular moment. Astronaut skin starts to affect the microbiome of the station, even as the station affects the skin of the astronauts."[25]

The microbiome of the ISS has a core group of over 55 microorganisms, thanks to more than 20 years of humans living in this tightly sealed, isolated environment.[26] During that time these microorganisms have been continuously exposed to the increased radiation levels associated with living in low Earth orbit. Some have speculated that this might result in changes in these microorganisms, potentially causing them to be more harmful. European Space Agency investigator Dr. Christine Moissl-Eichinger commented, "The idea we had was to find out whether the microbiome on board the space station is more resistant or more harmful than it would be on ground. Spaceflight causes some crew members to have periods of stress, and we wondered if the microbes would be stressed as well and might react in a bad way."[27]

Fortunately, the data showed that no "super-bugs" are developing on the space station! "We analyzed two sets of data, comparing data from the station with data from normal ground surroundings," said Moissl-Eichinger. "We found that microbial virulence was not increased, the resistances were not higher, the microbes were not stressed. We could not see any particular differences." The changing microbiome in the ISS simply seems to be the result of a constantly changing mix of people — and the cloud of bacteria, fungi and other microorganisms they bring with them — over the many years the station has been in operation.

As we push to explore farther into space, living for longer periods of time in habitats on the Moon or Mars, protecting other destinations in our solar system from contamination by microorganisms found on Earth is an important task. That critical job falls on the shoulders of the NASA Office of Planetary Protection. "It is always entertaining to see people's expressions when I introduce myself as the Planetary Protection Officer," commented NASA's Cassie Conley. "Most people think of the characters in the movie *Men in Black* when they hear this title . . . Actually, I was given dark Ray-Ban sunglasses my first day on the job."[28]

While the idea of a Planetary Protection Office sounds like something out of a science fiction movie, it is a serious undertaking. As we'll see in the next chapter, planetary protection attracts biologists and other scientists to avoid "forward" contamination of other worlds by terrestrial organisms and "backward" contamination of the Earth from any extraterrestrial life, if it exists.

CHAPTER 9

Interplanetary Protection

On November 17, 2019, a 55-year-old from Hubei Province in China inadvertently made history after being diagnosed with a new illness, a "novel coronavirus," marking the first documented COVID-19 infection in a human.[1] Since then it has spread through the world in a global pandemic unrivalled in the past century.

There are few places on Earth that have been spared — only four countries had no reported cases up to mid-December 2021. The rapidity with which it spread underscores the assertion that we live in a global village and what happens in one part of the world can rapidly affect the entire planet. When it comes to space travel, that problem becomes much larger: harmful microbes on Earth can easily be introduced into other worlds — and vice versa — with potentially devastating consequences. As humans explore the universe, it is our solemn responsibility to ensure the safety of extraterrestrial environments and of life here on Earth.

The concept of planetary protection from infectious disease dates back to the Apollo program, when NASA scientists did not know if even the lunar dust that the astronauts had been exposed to or the rock samples they were returning with might be contaminated with lunar microbes. A quarantine period was implemented after their return to protect the Earth from potential spaceborne microbes. Fortunately, there was no evidence of lunar organisms and the relieved astronauts left their 21-day quarantine to go home to their families.

Scientists haven't yet ruled out the possibility of other life in our own solar system, let alone in the universe at large. A NASA mission will fly to Jupiter's moon Europa for arrival in the 2030s, if the budget and schedule hold, to investigate habitability under the ice.[2] To be sure, the mission — called Europa Clipper — is not looking for the threatening seaborne creatures of *Europa Report*, a 2013 science fiction film that depicted an evil corporation sending hapless astronauts to the Moon to cope with a technical crisis and homicidal aliens, far from a rescue mission. Rather, Europa Clipper would be more likely to find "extremophiles," organisms that could survive in the harsh radiation environment near Jupiter. Although it may seem unlikely to find any signs of life in such an inhospitable place, the giant gas planet does provide enough energy to Europa (the moon) in its orbit that its ocean water isn't a frozen ball of ice. In fact, we have seen spurts of ocean water shooting from Europa's ice,[3] and the same is true of one of Saturn's moons, Enceladus.[4] Scientists are thus very interested in what may lie under the frozen surface of the numerous other "icy moons" that populate our solar system. And they are working hard to protect any life-forms that may inhabit them, including by throwing the Galileo spacecraft into Jupiter and the Cassini spacecraft into Saturn at the ends of their missions, on the off chance that either of these spacecraft carried harmful microbes, and to avoid the much smaller possibility that they could crash into the surface of an icy moon that may harbor organisms.

Then, of course, there is the perpetual hope for finding signs of alien life on Mars. Again, this isn't a search for Marvin the Martian or the homicidal creatures that populated H.G. Wells's *War of the Worlds* in 1897. What we know about Mars suggests that surface life would have a very tough time, given the planet has no atmosphere sufficient to shield its surface from radiation and little surface water outside of ice caps and frost. That said, beneath the planet's surface it appears there are abundant water deposits in which microbes could possibly exist,[5] evoking some comparisons with tiny Antarctic life that seems to be doing just fine, thank you, beneath half a mile of ice.[6]

As for the surface of Mars, the international Committee on Space Research has proposed designating certain areas as "Special Regions," meaning regions where conditions may be at least marginal for life to exist. For now, NASA and the European Space Agency have adopted strict contamination protocols to remove most chances of Earth microbes surviving on the surface of Mars. What will be curious to see are the planetary protection protocols of China, which landed the Zhurong rover on the surface on May 17, 2021.[7] China is only the second nation to soft-land anything on the Red Planet, and it will be interesting to see what, if anything, the country is doing to stop Earth microbes from getting onto the surface.

An example of strict anti-contamination measures is provided by Perseverance, NASA's newest Mars rover, which touched down on February 18, 2021. Perseverance is on the first stage of the agency's planned sample-return mission and its role will be to cache the most promising (ancient) signs of life for a future lander-rover to pick up and ferry back to Earth. NASA's website notes the rover and spacecraft were assembled in "clean rooms" equipped with powerful air filters, strong cleaners and stringent procedures such as using sterile cloths and alcohol wipes on hardware. (More durable pieces of the craft may be heated up to temperatures of 230 to 392 degrees Fahrenheit to bake any microbes into oblivion.) NASA strictly limited the number of spores allowed to go to Mars, and the launch hardware was also designed in such a way that no accidental rocket stages or similar debris would impact the surface.[8] These protocols evolve with every mission and NASA is already thinking ahead about what further safety measures to put in place for the sample-return mission.

And then there are the Planetary Protection Officer and the Office of Planetary Protection, whom we met in the previous chapter. In brief, NASA has established these important roles because it takes interplanetary safety very seriously. While most of their work and discussions are, for now, confined to the problem of keeping human microbes off other worlds, as humans travel farther around the solar system we are going to have to think carefully about keeping Earth safe from possible other life-forms. NASA's website lists the office's two main functions: "Carefully control forward contamination of other worlds by terrestrial organisms and organic materials carried by spacecraft," and, "Rigorously preclude backward contamination of Earth by extraterrestrial life or bioactive molecules in returned samples from habitable worlds."[9]

Representatives of the office, including the Planetary Protection Officer, work closely with mission managers to ensure the least possible threat to life as we move around other worlds. "To accomplish these goals, the Office of Planetary Protection assists in the construction of sterile (or low biological burden) spacecraft, the development of flight plans that protect planetary bodies of interest, the development of plans to protect the Earth from returned extraterrestrial samples, and the formulation and application of space policy as it applies to Planetary Protection," NASA states. The agency also seeks compliance not only with NASA policy, but with any international agreements with partners.[10]

There is ample scientific literature available on the endless number of ways in which a sample-return mission could proceed from Mars, but we will focus on the conclusions of one of the larger studies, published by NASA in 2002 and called "A Draft Test Protocol for Detecting Possible Biohazards in Martian Samples Returned to Earth."[11] This document will probably be updated with something much newer and more reflective of the latest technology as the sample-return mission timeline approaches, but for the time being, this 123-page document is a starting point to understand how one of the worldwide space agencies approaches the issue of keeping Earth safe from any microbes retrieved from Mars.

NASA opens the report by promising to abide by the U.S. National Research Council's recommendation on bringing Martian samples to our planet, which states: "Samples returned from Mars by spacecraft should be contained and treated as potentially hazardous until proven otherwise, and . . . rigorous physical, chemical, and biological analyses [should] confirm that there is no indication of the presence of any exogenous biological entity."[12]

The heart of properly isolating Martian samples to ensure human safety is a carefully designed and built Sample Receiving Facility, "combining technologies currently found in maximum containment microbiological laboratories . . . with those used in cleanrooms to preserve the pristine nature of rare samples," NASA states, but its next sentence identifies the problem: "Such an integrated facility is not currently available anywhere."[13]

Some hints for the facility may come from considering the thorny challenge of how we safely deal with infectious disease here on Earth. A strict protocol known as BSL-4 will be adopted, assuming a "high risk to the individual of aerosol-transmitted laboratory infection and life-threatening disease" until a good protocol is in place to show there is no need of such precautions.[14] The document also calls for a lot of research on decontaminating samples even before the mission lifts off, so that we have protocols ahead of time on how to deal with possible microbes. The report then enumerates various methods by which samples can be contained and decontaminated after being retrieved from the surface.

In April 2020, in the early weeks of the coronavirus pandemic, experts in a Space.com article pointed to two main issues to consider when planning the current sample-return mission. The first was that in 2019, NASA's Planetary Protection Independent Review Board recommended lowering the planetary protection category for much of the surface of Mars to Category II, which if adopted would relax the community's protocols in terms of overseeing or containing samples. The implication is it might be a lot easier to bring back samples from Mars than we feared, as much of the surface doesn't appear to harbor microbes. That said, how to protect Earth against microbes in the containment facility still hasn't been discussed fully, in large part because in 2020, the community anticipated that samples wouldn't come back for about 15 years.[15]

We also can't forget that Martian samples already *have* ventured to Earth in the form of various meteorites that made it to our planet. While we're careful here to downplay any claim that wee Martians made it to the surface of Earth, we must remember the lessons learned from past studies of those interplanetary rocks. One of the more famous or infamous meteorites, depending on your perspective, was ALH84001 — that's the one that NASA said in 1996 "strongly suggests primitive life may have existed on Mars more than 3.6 billion years ago,"[16] prompting an excited statement from then-president Bill Clinton about the result.[17] The research team identified what they perceived to be "the first organic molecules thought to be of Martian origin; several mineral features characteristic of biological activity; and possible microscopic fossils of primitive, bacteria-like organisms inside of an ancient Martian rock that fell to Earth as a meteorite."[18] The research was published in *Science*, but other teams quickly challenged the results, largely on grounds that the organic carbon compounds detected *can* indicate life, but they may also be signs of non-life processes.[19] The original research team has spent the ensuing decades doing more research in support of their claim,[20] which remains highly controversial.

The first few Apollo crews that went to the Moon, in 1969 and 1970, were pragmatic in their assessment of the mandatory quarantine when they returned to Earth. Indeed, the isolation "theater" was impressive; as soon as the first Moon-landing astronauts from Apollo 11 opened the door of the *Columbia* command module in the Pacific on July 24, 1969, they were thrown isolation garments, supervised from a distance while they scrubbed down, and then escorted (in the garments) to a little trailer known as the Mobile Quarantine Facility (MQF). Inside, they heard a speech from President Nixon and then, (still in the MQF, they were shipped to an isolated building at NASA's Johnson Space Center where they remained locked inside a secure facility with a support team for weeks, until the risk of lunar germs was deemed to have passed.[21]

Astronaut Michael Collins gave an interview decades later in which he revealed the flaw with this plan: "Suppose there were germs on the Moon. . . . [when] we come back, the command module is full of lunar germs. The command module lands in the Pacific Ocean, and what do they do? Open the hatch. You got to open the hatch! All the damn germs come out!"[22] Luckily for Earth, however, this major opportunity for lunar microbes to escape never resulted in a problem. By Apollo 14, the returning astronauts were down to just wearing masks, and by Apollo 15, no one worried about the problem of quarantine whatsoever — and with good reason, as several crews had already come back to Earth hale, healthy and not spreading a deadly virus among the community, showing that the stringent protocols were not required after all.

NASA described the quarantine in a brief history:

> In retrospect, it is possible to suggest that these concerns were overblown, but we only know that this is the case due to the careful preparations and the execution of the protective plans that evolved over several years in the buildup to the Apollo lunar landings. There were genuine concerns that the Earth's biosphere could be damaged by foreign life-forms or the lunar material itself with its chemical composition yet unknown. A July 1964 Back Contamination Conference staged between different U.S. governmental bodies also showed that the concerns extended to the legal sphere. The United States Public Health service had the duty to protect the citizens of the country against microbes, which might exist on the

moon. Similarly, the Department of Agriculture had the duty and the power to determine what kind of potentially biologically hazardous material entered the country — and in this particular case, the very planet itself. This made planetary protection not only a concern for NASA, but for the entire U.S. government.[23]

A new organization called the Interagency Committee on Back Contamination was formed prior to the first Apollo Moon-landing mission to determine quarantine protocols, and worked closely with NASA and Baylor University's College of Medicine in Houston to create a 571-page document determining the potential hazard of lunar material. While the bulk of media coverage focused on the astronauts, the organization's crowning effort was the Apollo Lunar Sample Return Container, usually referred to among lunar teams as "the rock box." At least two of these would fly on each mission, stored in an equipment unit on the base of the lunar module, for the astronauts to use on the surface. "To keep the lunar samples pristine, once they were closed on the lunar surface, nothing could get in and out of them, thanks to the sturdy construction, the locking mechanisms and a triple sealing mechanism at the lid."[24] The containers were rigorously designed "to withstand great physical forces, and even a computer simulation of a rock box was created to calculate the possible G-forces it might have to endure during the mission. It is perhaps a good indicator of the seriousness of this effort that they were constructed at the Y-12 Plant at the Oak Ridge National Laboratory in Tennessee, at a government facility best known for manufacturing nuclear bombs."

Reducing the risk of possible spread of harmful microbes from the Moon to Earth actually began while the astronauts were still on the lunar surface. Included in the Apollo 11 valve that emitted the lunar module's atmosphere on the Moon was a bacterial filter, just in case there was a threat. It turned out, however, that the larger problem was keeping the lunar dust from coming in. It would take several vacuumings to remove the dust as best as possible from the spacesuits and the equipment brought back from the surface. Although it turned out the dust was no real threat, it had a tendency to jam up and corrode equipment after long stays on the surface.[25]

When it came to ensuring the Apollo 11 astronauts' return to Earth would be safe for everyone, NASA knew the recovery plan was fraught. First there was the problem of trying to safely retrieve both a spacecraft and a crew from a potentially stormy sea. It hadn't been that long since an inadvertent door deploy caused NASA astronaut Gus Grissom's Liberty Bell 7 Mercury spacecraft to sink in the Atlantic in 1961, almost taking Grissom down with it (it was finally recovered in 1999, around Apollo 11's 30th anniversary). NASA also had to figure out how to minimize the risk of backward contamination and Earth's exposure to the lunar environment after the first Moon landing. At first, NASA contemplated leaving the Apollo 11 astronauts in the spacecraft for several hours, but figuring out how to keep the command module air-conditioned for several hours in the open Pacific Ocean was a large design challenge.[26] Hence the hatch was allowed to open, as much as the Apollo 11 astronauts later made fun of doing so for exposing the world to the potential risk from lunar microbes.

The astronauts' quarters in the Mobile Quarantine Facility (which you can visit at the National Air and Space Museum's Udvar-Hazy Center in Chantilly, Virginia) would have felt like a relatively roomy affair to a three-person crew who'd spent several days cooped up in the command module on the journey back from the Moon. The trailer was designed to keep six people housed for up to 10 days and, despite its relatively small size, included enough room for a lounge, a galley, a bathroom and of course areas to sleep. The best of 1960s knowledge and technology was deployed to isolate the inhabitants of the MQF from the outside world, NASA notes in a separate history: "Quarantine was assured in the MQF through the maintenance of negative internal pressure and by filtration of effluent air. Waste water from washing and showers was chemically treated and stored in special containers. Body wastes (urine and feces) were stored in special tanks . . . Items were passed in or out through a submersible transfer lock."[27]

After a long journey from the Pacific Ocean to Hawaii to Houston, the astronauts and their support personnel would finally be let loose into the Lunar Receiving Laboratory at NASA's Johnson Space Center in Houston. The laboratory was dozens of times larger than a typical house, with 83,000 square feet of space available; it not only included living quarters for the astronauts, but also administrative and support zones and ample space to examine the lunar samples.

"Special building systems were employed to maintain airflow into sample-handling areas and the Crew Reception Area to sterilize liquid waste and to incinerate contaminated air from the primary containment systems," NASA said. "The Lunar Receiving Laboratory was built to meet the most stringent biological containment requirements of the U.S. Army Biological Laboratories [at] Fort Detrick. This was a unique facility in many respects. It contained a vacuum chamber which permitted scientists to manipulate and examine lunar samples without breaking the vacuum or risking contamination of the samples or themselves. It had a low-level radiation counting facility and could safely accommodate a large variety of biological specimens."

The biological barriers for the returned Moon rocks and dust included procedures for a "vacuum complex," where the sample containers were opened, and biological cabinets, where the samples were tested. Outlining all the decontamination procedures would take a while, but some of the processes included sterilization with peracetic acid, the use of biological filters and negative air pressure to keep any possible germs contained, and strict procedures for handling samples, including the use of neoprene gloves.[28] Meanwhile, the astronauts (and all the personnel locked in the receiving laboratory with them) had daily medical examinations focused on recording their temperature, pulse rate and mental state. The crew provided biological samples on the 12th and 18th days of their quarantine before having a very thorough medical examination on the 21st day.[29]

We're still at a very early stage in terms of considering all the quarantine protocols that astronauts will face after coming home from a Mars mission, but we can see how NASA has updated its protocols in the decades since Apollo. The agency faced a few situations where astronauts undertook Apollo missions with illness, including the well-documented incidents of head colds among the Apollo 7 Earth-orbiting crew in 1968, an illness of commander Frank Borman during the first Moon-orbiting mission (Apollo 8) of 1968, and of course Apollo 13 of 1970. While the 1995 movie portraying the perilous mission played up the drama among the astronauts, Apollo 13 was famous for switching out a member of the prime crew (Ken Mattingly) for a member of the backup crew (Jack Swigert) just days before launch, after Mattingly was unwittingly exposed to the German measles by another backup-crew astronaut, Charlie Duke.

Duke's three-year-old son had the illness and NASA flight doctors determined most of the prime crew and backup crew had immunity — except Mattingly, who was ordered off the mission.

"Dr. Charles A. Berry . . . was concerned that if Mattingly were to come down with German measles, given the virus's incubation period it would likely be around the time they were in lunar orbit, possibly while he was alone in the CM [command module]," NASA stated. "Symptoms such as fever, rash and headache could interfere with his ability to perform intricate maneuvers to rendezvous with [crewmates] Lovell and Haise in the LM [lunar module] as they returned from the moon."[30]

While Mattingly ultimately never came down with the illness, the last-minute disruption to training added to strains the astronauts later encountered en route to the Moon, when a series of technical problems led to an oxygen tank rupturing in space and severely damaging the command module.[31] The astronauts, happily, made it home safely due to a combination of sheer grit and leadership displayed not only by the crew, but by mission control and their supporters in Houston — a story we covered in our previous book, *Leadership Moments from NASA*.

In any case, these various incidents showed NASA that quarantine protocols prior to flight might be helpful for isolating crews, and to this day the agency is very careful to limit contacts in the days and weeks leading up to flight. (These protocols were even more tightly restricted during the coronavirus pandemic, but we will focus on the more "normal" procedures as the COVID-19 precautions might ease by the time this book is published.) NASA's "Flight Crew Health Stabilization Program" document on the NASA Technical Reports Server provides a brief glimpse at what astronauts undergo before and after flights:[32]

> The primary purpose of the Flight Crew Health Stabilization Program (HSP) is to mitigate the risk of occurrence of infectious disease among astronaut flight crews in the immediate preflight period. Infectious diseases are contracted through direct person-to-person contact, and through contact with infectious material in the environment. The HSP establishes several controls to minimize crew exposure to infectious agents. The HSP provides a quarantine environment for

the crew that minimizes contact with potentially infectious material. The HSP also limits the number of individuals who come in close contact with the crew. The infection-carrying potential of these primary contacts (PCs) is minimized by educating them in ways to avoid infections and avoiding contact with the crew if they are or may be sick. The transmission of some infectious diseases can be greatly curtailed by vaccinations.[33]

NASA normally starts the quarantine at least seven days prior to launch, which is a typical incubation time for a minor illness such as a cold. After the onset of the global COVID-19 pandemic the preflight quarantine was increased to fourteen days. The crew commander may ask for an earlier quarantine time to help with the crew shifting their body schedules for a night launch, or some other unusual situation that can affect circadian rhythms.[34]

The quarantine facilities vary depending on where the astronauts are launching; as of 2021 it was possible for NASA astronauts to launch from either the Kennedy Space Center in Florida or Baikonur Cosmodrome in Kazakhstan, subject to mission requirements for launch windows as well as the schedule for the International Space Station (the ISS requires a minimum crew at all times). That said, a typical quarantine facility is supposed to include adaptive lighting for "sleep shifting"; individual bedrooms and bathrooms with their own lighting and climate controls; crew training and briefing rooms allowing the astronauts to interact with the world through computer, telephone or videoconferencing; a medical examination room; common dining and entertainment areas; and common exercise facilities.[35]

The list of people approved to enter the facility is restricted to "personal contacts," who must avoid such normal activities as shaking hands; if they must touch the crew members (medical personnel, for example), they must use soap and water or hand sanitizer.[36] Extended family, guests and VIPs may attend as long as they are cleared by a brief medical examination.[37]

Naturally, a quarantine period is also implemented after crew members get home. NASA astronaut Victor Glover, for example, tweeted in May 2021 that the members of SpaceX Crew-1 waited a few days after landing to receive their COVID-19 vaccine. "Had to give the immune system a few days to adapt before getting vaccinated," he explained.[38]

All these protocols have allowed astronauts to fly to space largely free of communicable disease for several decades, although as we have shown elsewhere, spaceflyers face other medical complaints not related to quarantine upon reaching orbit. With regards to planning for a far-off Mars mission by humans, the National Academies of Sciences, Engineering and Medicine reviewed a report in 2020 by NASA's Planetary Protection Independent Review Board.[39]

"With the probability of humans landing on Mars ever more realistic, our reports recommend that NASA conduct research to see if there can be a Martian 'exploration zone' where humans can land and contamination, if it occurs, would do no harm," said Scott Hubbard, an adjunct professor of aeronautics and astronautics at Stanford University, who coauthored the NASA report. Hubbard, also the former director of NASA's Ames Research Center in California, added that spacesuits can possibly leak, allowing microbes from humans to irreversibly contaminate the surface of Mars.[40]

Beyond the Mars surface exploration problem, the report points to other emerging threats from human space activities. SpaceX once sent a Tesla car, complete with a spacesuited mannequin affectionately named "Starman," to the vicinity of Mars. As much as this activity spurred worldwide public interest in space exploration, the report warns that missions such as this should be subject to quarantines as strict as those for typical NASA missions leaving Earth.[41] Even smaller spacecraft, Hubbard said, should be considered in these discussions. "Small spacecraft with the potential to go to deep space are being developed at very low cost at both universities and companies, and we highlighted concern about whether these small spacecraft will be overly burdened by the cost of PP [planetary protection] requirements," Hubbard said.[42]

The history of life on Earth is largely microbial, making up roughly three billion of the four billion years life has been present on our planet. The search for more advanced extraterrestrial life-forms elsewhere in our solar system and beyond is compelling and has captured our imagination; but it is more likely that life on other worlds will also consist of small microorganisms suited to live in extreme, harsh environments. Wouldn't it be ironic if the first evidence of life on Mars was microorganisms brought from Earth on the rovers designed to search for signs of life on the Red Planet? NASA has been extremely vigilant in controlling this risk, but once humans begin exploring the surface of Mars, Earthborne bacteria and other microorganisms will travel with them on their skin and in their digestive systems. The containment lessons learned from Apollo will be important in reducing the probability of lunar microbial colonization and will undoubtedly be improved for the first human missions to Mars. Either way, planetary protection at home and abroad is here to stay.

CHAPTER 10
Touching the Moon

R eferred to by *the New York Times* as "one of the great clinical writers of the 20th century,"[1] Oliver Sacks was the consummate academic physician. His compassion, empathy and curiosity permeate his descriptions of unique afflictions of the nervous system, challenging us to think about who we are and how we interact with the world around us. Of his many stories, "The Disembodied Lady" best reveals the importance of proprioception; often called the sixth sense, this is what enables us to know the position of our limbs and body.

Perhaps the reason most have never head of proprioception is our tendency to take it for granted. Sight, smell, taste, hearing and touch provide a powerful source of sensory information about the world around us, enriching our lives with color, texture and music. The acute sensations we perceive and experience through our senses bring heightened awareness to our lives, connection to others and the world around us, and a feeling of wonder. Limb position, however, is a little more abstract. We don't typically think about where our limbs are or value that sensation; that information is processed and interpreted in the background without us being conscious of it happening. That is, until an extremely rare event occurs and our sixth sense fails.

Sacks tells the story of Christina, a 27-year-old mother admitted to hospital for routine gall bladder surgery. Before the operation could take place, her gait became unsteady; her movements, no longer finely coordinated, had become awkward, flailing gestures. She was dropping things without explanation. The surgeons asked their neurological colleagues for a consult. It was a profoundly upsetting experience for her. "Something awful has happened," she said to Dr. Sacks and his neurology resident. "I can't feel my body. I feel weird — disembodied."[2]

Sacks explained to her that our awareness of the body comes from vision, balance and proprioception, a term first used by Sir Charles Sherrington, who won the Nobel Prize in Physiology or Medicine in 1932.[3] It relies on the brain integrating signals from sensory receptors in the muscles, tendons, joints and skin that respond to changes in muscle length and force, to joint rotation and to local bending of the skin.[4] Without it, the body's innate ability to sense itself is lost.

In most cases, loss of one sensory input causes a greater reliance on others. Visual impairment is offset to a degree by increased reliance on hearing, smell and touch. Perhaps the same might be true with loss of proprioception. Christina listened to Dr. Sacks explaining where the sense of body comes from and concluded, "What I must do then is use vision, use my eyes, in every situation where I used — what do you call it — proprioception? I've already noticed that I might 'lose' my arms. I think they're in one place, and they're in another. This 'proprioception' is like the eyes of the body, the way the body sees itself . . . My body can't 'see' itself if it's lost its eyes, right? So I have to watch — be its eyes."[5]

Ian Waterman also lost his sense of proprioception. At the time it happened, he became one of 15 people in the world unable to sense where their limbs are and how they're moving.[6] He recalled, "[After] what I thought was flu, and [I] gradually got weaker and weaker and weaker, and was eventually taken into hospital and woke up the following morning . . . I still had movement, I could move an arm and I could move a leg, but I had no very fine control over that. The consequences were that I couldn't control my body without looking at what was happening. It wasn't for some time that I found that I had to associate vision with picking up feedback."

Activities that normally take place without the need for continuous monitoring became a challenge. Waterman must continually look at his feet and the path ahead when he walks. "When you're walking down the road, you don't think about every step that you're taking. I do, otherwise I lose touch with where things are . . . I walk in the way that I do, which is inclined

forward and looking down . . . my whole body is monitored and controlled."
Something that was effortless before has become a demanding task.

Integrating data from multiple sensory inputs is an important element of the feedback needed for fine motor control — performing precision tasks accurately. Mismatched multisensory inputs can create challenges for the brain to integrate, which can lead to misperceptions or illusions.

An interesting example is the Pinocchio illusion, first described by Brandeis University Professor James Lackner in 1988.[7] The illusion is created by giving a subject conflicting sensory inputs. Seated behind another person, the subject is asked to extend their arm over the other person's shoulder and told that it will be moved by the observer to tap the end of the other person's nose. The observer asks the subject to close their eyes and then moves the index finger of the subject's extended arm to touch the nose of the other person while touching the subject's nose at the same time. The unexpected but synchronous tapping of the subject's nose gives them the illusion that they are touching their own nose, but it is now farther away — hence the Pinocchio name.

University of Barcelona researcher Dr. Mel Slater explained this phenomenon: "The brain has a contradiction to resolve, which is how come my hand is out there and it feels like I'm tapping a nose . . . at the same time, my nose is feeling tapped. The brain doesn't like contradictions, it likes solutions, and the solution it comes up with [is] — okay so my nose is really that long, and that's as simple as it is."[8]

The body has developed these mechanisms to move and perform tasks gracefully in a gravitational world. What happens to astronauts living and working in space? Would the ability to perform fine motor tasks degrade in the absence of gravity? Whether working and living inside the International Space Station or performing a spacewalk in the vacuum of space, dexterity is just as important as it is on Earth. Researchers performed experiments on the NASA Neurolab (STS-90) mission in 1998 to understand more about what happens to human proprioception in space.

Scientist Otmar Bock and his team applied several tests before, during and after that space shuttle mission to assess changes in the visual-motor performance of astronauts.[9] They developed a compact research tool called the Visual-Motor Coordination Facility, which used an electronic target display to study the accuracy and speed of performing grasping, pointing and tracking tasks in the absence of gravity. The tool was designed to test the importance of visual feedback during motor tasks by allowing subjects to either view their hand or not, based on the different experiment protocols.

The pointing task required subjects to point with their index finger from a common starting position to targets displayed six centimeters (2.4 inches) to the right, left, above or below. In the grasping task, subjects gripped luminous discs between the thumb and index finger, with the displayed discs located away from the starting point in a randomly varying manner. The tracking task had subjects use their index finger to follow a target moving at constant speed along a circular path.[10]

When visual feedback was removed by not allowing subjects to see their hand during some tasks, the astronauts had to rely primarily on proprioceptive cues to perform the task. Would the sense of limb position be degraded due to a mismatch between visual and proprioceptive cues? Would proprioceptive senses differ in the absence of gravity, resulting in reduced accuracy and speed of performance?

The tests in space showed that the accuracy of the inflight pointing and grasping tasks did not change when the subject's hand was not visible, but there was slowing of task performance in both cases. There was no difference in speed performing the tracking task or with reaction times assessed during the task. However, it was noted that without visual feedback the subjects did not accurately follow the circular track of the target; the recorded tracks showed more of an ellipsoid shape than a circle. Researchers found that the orientation of the ellipsoid shape was inclined from the vertical by 20 degrees in sessions before and after the mission, while during the mission the responses were less inclined in the absence of gravity.[11]

The slower inflight task performance observed in this experiment had previously been described by other researchers and had been attributed to greater reliance on visual feedback to perform the task. Bock and his team showed that couldn't be the case, as similar results were obtained when visual monitoring of task performance was removed. They proposed that "response slowing is a symptom of adaptive restructuring in the brain." In essence, the brain appeared to be reprogramming the way it was processing the information when it lacked visual clues.

The reduced inclination of the ellipsoid tracking data during the mission was attributed to a "stronger dependence on an egocentric [self-centered] reference frame" based on proprioceptive inputs. In other words, it appeared that the astronauts were influenced more by their perceived orientation of up and down in tracking the objects and were less influenced by external cues provided by gravity. In a gravitational environment, tracking the target required astronauts to hold their arms up and away from their sides against the force of gravity. While not particularly onerous, over the 12 to 16 seconds of the task it is possible that the subject's elbow might drift downward, creating an inclined ellipsoid shape. Without gravity that influence disappeared.

While humans are used to working against the force of gravity on Earth, pilots have anecdotally reported challenges in accurately reaching out to cockpit switches during periods of high G-forces experienced when flying in military operations, aerobatics and during the liftoff phase of the space shuttle. Performing precise time-critical tasks is important for both pilots and astronauts. Perhaps the slower inflight task performance observed in Bock's experiment might have had an impact on the ability of shuttle commanders to provide timely, accurate inputs while landing the space shuttle, and further studies might be helpful to understand the risks associated with manual spacecraft landings after long-duration spaceflights.

As a species we are remarkable in our ability to adapt to the circumstances we're in, and it is likely that astronauts participating in future missions will quickly adapt to the microgravity environment of space, or to the partial gravitational worlds on the Moon and Mars, and readapt to the Earth's gravity when they return. However, the experiences of Christine and Ian, along with the findings from scientific studies, suggest that there are many subtleties to proprioceptive disorders that may affect astronauts as well as children and aging individuals.

There is a growing body of clinical information describing three different categories of sensory processing disorders (SPD) in children.[12] One of the categories is called sensory-based motor disorder and has two subtypes: one a postural disorder and the other a challenge with coordination. The term "slumper" has been used to describe kids with the postural disorder: "The slumper has difficulty with movement, and moves in a clumsy, disorganized way. [They] may have difficulty stabilizing [themselves]. [They] may struggle to run without tripping over [their] feet. Kids with postural disorder have difficulty crossing the midline [with their limbs] or using the hands and feet [from] one side of the body on the other side."[13]

The second subtype is called dyspraxia, also known as clumsy child syndrome. These children tend to be "fumblers" and have difficulty planning and executing actions like throwing or catching a ball or frisbee. They prefer toys, games and situations that are familiar, and tend to avoid new situations that require motor planning. The intent of researchers and clinicians is not to label children, but to develop strategies that might help them. Based on the knowledge of how our sixth sense works, potential solutions might include using multisensory teaching and perceptual motor training to help them learn to use several different senses to perform fine and gross motor tasks. One of the most effective strategies for helping children with SPD is the use of occupational therapy that incorporates sensory integration techniques.

A simple example from the world of dance shows us how learning new ways of doing things can help children with SPD. Surprisingly, dancers known for their grace during a performance have sometimes been known to be clumsy in day-to-day activities.

"I definitely think the adaptations your body takes on when being a full-time competitive or professional dancer can make you more clumsy if you're not mindful," said Monika Volkmar, a strength coach and owner of the Dance Training project.[14] For instance, audiences admire the pointed feet of skilled dancers, but continually pointing your feet during movement is not natural, noted Volkmar.

"When you swing your foot through when walking, it's more likely you'll stub your toe on the ground as compared to someone with ample dorsiflexion [lifting the foot off the ground]. To compensate, dancers tend to walk with their feet pointing more out to the side to allow for their foot to clear the ground and not trip over it." She suggests that cross-training by practicing other activities like martial arts, yoga or resistance training can help dancers learn to move smoothly when "not in dance mode."

For those of us who are not professional dancers, recreational dancing is not only an excellent way to relax, but it can also "significantly improve muscular strength and endurance, balance, and other aspects of functional fitness in older adults."[15] Even Apollo 11 astronaut Buzz Aldrin joined in the fun on the TV program *Dancing with the Stars*. In an interview with Space.com, Aldrin commented, "I'm not known for that agility. I'm one of those people who need continued exposure. Whether it's flying a spacecraft or trying something out on the dance floor, it takes a while to really get comfortable with it."[16]

Other movement training strategies help maintain our balance and gait as we age. Tai chi has been referred to as one of the most "exciting" interventions for seniors because it benefits balance and mobility, and provides a sense of calm in its practitioners. A *Harvard Gazette* article described it as "improving balance, flexibility, and mental agility, [and] it also reduces falls, the largest preventable cause of death and injury among older adults."[17] In an interview for the article, 84-year-old Elaine Seidenberg said, "In class, we wave hands like clouds. And after class, we walk on clouds . . . When I come out, I feel at peace with myself and the world. Somehow when we age, we become less coordinated and a bit more clumsy, but I feel more graceful."

Like imitating or watching clouds, there is an inherent tranquility associated with graceful movement in the weightlessness of space. Some astronauts are aptly described as a "bull in a china shop" until they become used to moving slowly with only fingertip pressure to propel them from one location to another. Others adapt more rapidly to this unique environment and seem born to move with fluidity inside the spacecraft. For veteran astronauts, their bodies remember what it is like to live in space — something that is evident to the first-time flyers as they watch their experienced crewmates slowly undo their harnesses and float gracefully out of their seats after the harsh, high-G ride to space. Current evidence suggests that astronauts will adapt to smoothly performing complex motor tasks, gross and fine, whether in microgravity or in a partial gravitational environment — as they reach out to touch the Moon once again.

CHAPTER 11
Reversible Aging

In 1995, Senator John Glenn was paging through a book called *Space Physiology and Medicine*, published two years before. Then 73 years old, he spotted a chart that captured his imagination. Listed were 52 kinds of physical changes routinely experienced in space by orbiting astronauts, including osteoporosis, cardiovascular difficulties and alterations in the distribution of fluid in their bodies. Glenn, who had spent years on the Special Committee for Aging, realized that he was looking at a list of complaints common to seniors[1] — yet the typical astronaut was in their thirties, forties or fifties. Given the similarities between the reversible changes seen in astronauts and those associated with aging, Glenn wondered if there were any plans to fly senior citizens into space and called the authors of the publication along with other space doctors to find out. To his surprise, despite shuttle missions going up every few months, nobody knew of any plans to send an elderly person into space.[2]

"I wondered why the science had to wait," Glenn wrote in his autobiography. "Shuttle flights were going up regularly. Why couldn't room be made on one of them for some experiments on senior citizens? And then I began to think, 'Why not me?'"[3]

Early in his career, Glenn trained as a marine and was an accomplished test pilot, breaking numerous flying records and even appearing on television before being scooped up by NASA in the very first astronaut recruitment drive. Famous for his clean-cut image and charm with the press, he was a favorite to be the first to fly in space. Indeed, he was on NASA's first orbital mission, Friendship 7. But his plans for a return to space were stymied by the Kennedy administration, who deemed Glenn too valuable a national hero to return to orbit.[4] So Glenn left NASA and moved into politics, never expecting that the chance for space travel would come up again.

Several things played into Glenn's favor as he lobbied NASA and then-administrator Dan Goldin in 1995 for a flight opportunity. NASA was, of course, wary of sending a senior into space — let alone one as famous as Glenn, who had been a Senate politician for decades. Yet NASA did have 42 years of data on the astronaut,[5] who had consented to regular medical checks under a program the agency has for retired astronauts. Glenn was also in remarkably good health; one media report described the astronaut, shortly before his flight at age 77, as "one very fit, lifelong jet jockey who power-walks two miles a day on Earth and has passed all his preflight physical tests."[6]

Glenn, by far the oldest person to fly into space during that time, defended critics who said that flying only one senior into space (particularly a remarkably fit one) would not provide any meaningful information, given that ideally you would want a sample of around 10 to 12 people of different genders, fitness levels and the like to learn if the aging effects of space travel on astronauts are similar in older people. "Whether information collected from a sole geriatric volunteer will be sufficient to be meaningful is contro-versial," a reporter wrote.[7]

Glenn also told NASA, in a 1997 oral history just prior to his flight, that this was not simply a bid to get himself one last joyride in space, pointing out that countries around the world were wrestling with the issue of aging populations and the responsibilities of taking care of them. His hope was that spaceflight would spark more insights. "This is something to really start a new area of research that I think can be very, very important, and that's the reason it's so fascinating to me. Much as I'd like to go up again and just joyride around, we don't have the luxury in our spacecraft yet of just letting people go up just to get the view. This is an area of very, very proper research that has the potential to it of enormous benefit for — well, the graying of nations, they call it, all over the world."[8]

At 77, Glenn flew as a payload specialist on the STS-95 Spacehab mission, meaning he was responsible for some of the many life science and microgravity experiments on board. All reports indicate that he was careful to contribute; a media report said he "received good reviews for his work ethic and affability"[9] and didn't bother himself with the attendant celebrity that his name could have brought. On board the mission, he made sure to fulfill crew responsibilities with not just experiments but also housekeeping and meal preparation.

Glenn was the first elderly person to have flown in space; previously, the oldest astronaut to participate in orbital spaceflight was Story Musgrave, a highly regarded space shuttle veteran who made his last venture into space at the relatively youthful age of 61. But that has all changed with the recent suborbital flights of Wally Funk at age 82 and William Shatner at 90 on Jeff Bezos's Blue Origin spacecraft. Shatner now holds the record as the oldest person to have reached space. (Incidentally, as of 2020 the well-regarded collectSPACE forum lists five active NASA astronauts in their sixties.[10])

Government astronauts have by far been the majority of the almost 600 individuals who have flown to space since 1961. In the past a few private tourists flew aloft, mainly associated with the Mir space station and Russian International Space Station activities, but these were largely rich entrepreneurs in their forties or fifties. Yet the coming years will see a bonanza of private activity and travellers of all ages. As of 2021, missions such as Inspiration4, Axiom Space and dearMoon are early ventures by private companies planning to bring nonprofessional astronauts into space and, in some cases, on to the International Space Station. Meanwhile, SpaceX, Virgin Galactic and Blue Origin have all flown civilians on suborbital flights or to orbit. It may be that these kinds of missions will bring individuals of various genders, abilities and ages into space — making it all the more important that we begin to understand the effects of microgravity on a wider range of people.

Glenn performed several experiments on himself in space, sometimes with the assistance of fellow astronaut Scott Parazynski, a medical doctor who was on board with him. Metabolism experiments had him swallowing one kind of amino acid in pill form and another kind by injection, as proxies for hormonal changes caused by muscle atrophy or protein loss. X-ray studies recorded changes in lean body mass. Glenn also charted his sleep patterns on four nights, comparing the results of taking melatonin or placebo pills, and tracked his core body temperature with a mini-thermometer embedded in a capsule he had swallowed. His brain waves were monitored using devices clamped to his scalp, wrist and chest and, like many other astronauts, Glenn also gave samples of blood and waste for analysis.[11] While the results of these experiments appear to have been largely aggregated with those of other astronauts, one of the sleep tests noted that there didn't seem to be any real difference between the older astronaut and his much younger colleagues.[12]

What Glenn's flight showed was that "aged" does not necessarily mean "infirm" — and that should give hope not only to seniors but also to the much younger astronauts who typically spend months or even a year on the International Space Station and experience some reversible symptoms of aging as a result. Space temporarily induces many of the effects of aging even in younger crew members, as we will explain, but it appears these effects are largely reversible over time once the astronaut reaches the ground. Still, the link between microgravity and aging is so strong that some researchers are trying to apply the results to find countermeasures to help earthbound seniors as well as astronauts. Among these researchers is Richard Hughson, the Schlegel Research Chair in Vascular Aging and Brain Health at the Schlegel–UW Research Institute for Aging in Ontario, Canada.

Hughson, who has been working on this line of research for many years, found that "during a six-month space mission, an astronaut's cardiovascular system can age by up to 10 or 20 years." Since 2007, he has spearheaded a series of four Canadian experiments to investigate in detail why this is the case and what countermeasures can be put in place to prevent and reverse the sudden aging of astronauts in space. The first study, which ran from 2007 to 2010 and was called Cardiovascular and Cerebrovascular Control on Return from the International Space Station, examined how the hearts and blood vessels of astronauts alter due to spaceflight.

"The absence of gravity in space disrupts the normal circulation and distribution of blood inside the body," the Canadian Space Agency (CSA) said in a fact sheet[13] about the experiment. "Astronauts experience puffy faces and 'bird legs' as blood moves from the lower body and congests in the head and chest," as we described in Chapter 2.

"When astronauts return to Earth, the distribution reverses and blood collects in their lower body," the CSA added. "If not enough oxygen-rich blood reaches the brain, some astronauts could experience dizziness or blurry vision, while others might even faint."

The study followed six astronauts who strapped on devices that examined their heart rate, blood pressure and physical activity in several 24-hour periods before, during and after their mission. This allowed the investigation team to see how the system changes over time. Naturally, there was individual variation, but the study concluded that more exercise and other countermeasures — for example, wearing devices to alter blood flow — would be helpful for astronauts to recover; this research was also relevant for seniors, who are facing similar challenges because of their age.[14] (Uniquely, Hughson's control experiments on Earth include studies at a seniors residence that is closely associated with the University of Waterloo, allowing him to test consenting seniors in situ without the bother of them needing to commute to a lab.[15])

The success of this study then spurred the Vascular Series, which comprises three sets of experiments to date. The first one, Vascular, looked at changes in the blood vessels and the heart over time, specifically examining how arteries change and perhaps stiffen as blood pressure alters due to less physical activity in space. The artery stiffening then came under even more scrutiny in Vascular Echo, which used the countermeasure of having astronauts wear a leg cuff to reduce how much blood is in the upper body and shift it to the lower body, better mimicking proper blood flow on Earth. Vascular Echo also investigated whether blood flow after exercise is affected by spaceflight, and sought to monitor astronauts for · a year postflight to see how the body alters after a long-duration stay on the space station. The latest in the series, Vascular Aging, looks at how and when insulin resistance may happen during a space mission; it also examines how radiation exposure may affect cardiovascular health and, like Vascular Echo, explores how well an astronaut recovers after coming back to Earth.

Each study in the Vascular Series has nine participating astronauts, making more than 20 who have participated in the larger study so far. That's no small number of ISS astronauts; before commercial crew was available to increase the size of crews, there would be only about six eligible people working on the U.S. side a year (assuming one launch of three cosmonauts or astronauts each quarter of the year, for a total of 12 people on the ISS.

Only non-Russians participate in these NASA studies, and Russians tend to make up half of ISS crews).

One drawback to the small sample sizes in space is the amount of time it takes to collect and then parse enough data, but we do have some preliminary results from Vascular, which found that arteries of astronauts tend to stiffen by 17 to 30 percent. Astronauts also may be at more risk of type 2 diabetes because they develop insulin resistance and have difficulties processing glucose. Female and male astronauts alike are affected by insulin resistance, although males tend to have more of an issue. Vascular's data was collected between October 2009 and March 2014. We're still awaiting the results of Vascular Echo, which wrapped up in 2021 after five years of study. Vascular Aging just began collecting data in 2019 and will finish in 2024, according to the CSA.[16]

"We've looked at the carotid artery in particular because it changes a lot with space flight due to the physical inactivity and due to a change in pressure gradient," Hughson told the Canadian Institutes of Health Research in a 2015 interview about Vascular. "What we found is that we are seeing astronauts coming back from space with carotid arteries that have aged the equivalent to 20 to 30 years in stiffness. It is very comparable to what you would see in an aging population. So we think the method or the mechanisms are quite different in some ways but there are some similarities too."[17]

For seniors, the implications cannot be clearer: there is more that must be done as people age to counter the effects of staying sedentary. "Cardiovascular disease is the number one cause of death on Earth. According to the Heart and Stroke Foundation, nine out of 10 Canadians have at least one risk factor for heart disease or stroke, including lack of physical activity," the CSA said in a fact sheet.[18] "Vascular's results suggest that a daily session of aerobic . . . exercise alone is not sufficient to counteract the effects of an otherwise sedentary lifestyle."

Hughson underlined the need not only for exercise, but for consistent exercise. Many doctors have pointed to the risk of sedentary work for industrial populations in general; in space, which mimics the effects of aging, finding consistent ways to stay active appears to be even more urgent.

As you can imagine, astronauts don't have to really work their lungs, legs and feet to move from place to place in space. Their lungs, without proper exercise, become deconditioned. As we have seen, their soles even become soft from disuse as they float around. Conversely, astronauts tend to become very adept at using their arms and their core to move around; they even learn how to push off from one wall to the next with the simple flick of a finger. On Earth, they'd be next to helpless upon landing without countermeasures in space that include a tailored exercise regime, which is why studies into how to assist the cardiovascular system are so crucial.

"One of the things we've identified with the astronauts is that they're really doing only about 30 minutes per day of aerobic type of exercise and the rest of the day they're floating around. So clearly 30 minutes per day is not enough exercise," Hughson said.[19] "Astronauts, for sure, definitely need to be moving more. We know from many studies here on Earth that physical activity is a good thing in many, many ways for your body, your cardiovascular system, your metabolic control system and muscles and bones and joints, so physical activity is definitely good for us."

An astronaut returning to Earth arrives much like a delicate senior who may have several health issues preventing them from moving for long periods of time. The aim, for the first few months or so back on Earth, is to get the astronaut back to a typical state of middle age — able to drive, to exercise, to actively participate in work and home activities and to continue contributing to research and other missions at NASA.

Astronaut Leroy Chiao once wrote about the effects of coming back to Earth after short space shuttle missions of roughly 10 to 14 days:

> The return is dramatic. Your balance system is turned upside down, and you feel very dizzy. When you stand up for the first time, you feel about five times heavier than you expect. All of this can be unsettling, and nausea is not unusual. After a long-duration flight of six or more months, the symptoms are somewhat more intense. If you've been on a short flight, you feel better after a day or two. But after a long flight, it usually takes a week, or several, before you feel like you're back to normal. All you want to do is lie around, because in that position, you are not dizzy. But, in order to recover your balance to the point of being able to exercise, you must force yourself to walk around; it is this physical activity that really accelerates your recovery.[20]

After Christina Koch spent a year in space, NASA wrote a feature suggesting 10 ways in which the healthy 40-something would need to recover when coming back to Earth. For example, coming back means that sleep will no longer be simple floating bliss, that smells will suddenly be frequent, and that sensory inputs like wind on the face will suddenly happen again — much like what a convalescent senior will feel when coming outside after a long illness or at the end of a long, cold winter. But weight and balance are also things that any senior or any astronaut needs to take into account.[21]

"On Earth we rely on our eyes and inner ear to maintain stability. In orbit, without gravity pulling down, the mind quickly stops listening to the inner ear. The eyes take over . . . we rely solely on visual cues," Koch said in the article. "From what I've been told, it takes a couple days after landing for the mind to start listening again. The human body's ability to adapt to its environment is nothing less than impressive."[22]

NASA astronaut Jessica Meir, who landed on Earth as the coronavirus pandemic's first wave was gripping the world, found herself facing border closures and longer quarantines during the standard two-day journey from Kazakhstan to Houston. Back at home, Meir isolated herself for a week instead of the usual two days. "Something about that spaceflight environment does have a direct influence on our immune system, and that's why they wanted to be extra conservative with what we were exposed to first upon coming back," Meir told NBC News.[23]

The recovery from space actually begins as astronauts dress for the ride home. Both NASA and Roscosmos, the Russian space agency, use garments with lower-body compression to assist with adjusting fluid shift. NASA keeps refining the design by studying test subjects in a laboratory following a period of bed rest, and of course monitoring the effect on its own astronauts after spaceflight. NASA notes that the garment also has applications for people on Earth who have a tendency to faint or experience difficulties in blood pressure.[24]

After the astronaut gets home, aches and other muscle-related problems are common. We use different muscles for floating around in space and often don't need to be using supports in our neck and body to keep us upright, unless exercising, the Canadian Space Agency said in a brief explainer on countermeasures. "On Earth, we must constantly use certain muscles to support ourselves against the force of gravity. These muscles, commonly called antigravity muscles, include the gastrocnemius (calf muscles), the quadriceps (thigh muscles), and the muscles of the back and neck. Without regular use and exercise, our muscles weaken and deteriorate."[25]

Even before an astronaut goes to space, NASA establishes a baseline for each person to customize their rehabilitation program for their return. In other words, they make sure to tailor the training so that you are at a level of capability after the recovery period that is similar to your fitness before travelling to space, rather than aiming for an idealized standard that may not be a fit for all bodies. The rehabilitation period is lengthy and numerous astronauts have pointed to the need to take it easy and to get family members assisting during this delicate time, which can last months.

Fortunately, the astronaut strength, conditioning and reconditioning (ASCR) specialists assigned to help each astronaut with his or her recovery have more than 20 years of experience working with ISS crew and know what sorts of countermeasures are required to keep the returning spaceflyers safe and healthy. "The postflight reconditioning program is designed to stress the body systems that affect the following: aerobic capacity, muscular strength, power, endurance, stamina, bone, balance, agility, coordination, orthostatic tolerances, proprioception, neurovestibular function and flexibility," read a postflight reconditioning brief on NASA's Technical Reports Server, from 2011.[26]

"Postflight reconditioning begins on landing day, is scheduled for two hours per day, seven days a week for 45 days and is tailored to the specific needs of the astronaut. Initially the program focuses on basic ambulation, cardiovascular endurance, strength, balance, flexibility and proprioception. The program advances through 45 days and specific attention is given to each astronaut's overall condition, testing results, medical status, and assigned duties after their mission."

A 2017 feature in *Men's Journal* described the recovery Robert "Shane" Kimbrough was about to go through on his imminent return from space after commanding Expedition 50 on the ISS. Kimbrough is no athletic slouch. Even though he was pushing 50 at the time of the interview, he was described as "energetic" and "military-fit," and shown doing a rigorous cardio and weight lifting routine that he had maintained for two years to get ready for space. Despite Kimbrough's military experience, though, lead

ASCR Mark Guilliams had some helpful suggestions to assist Kimbrough with his exercises. "He came from a military background and hadn't done a lot of weight training," Guilliams said in an interview with the magazine. "We taught him the squat, the deadlift."[27] He added:

> Most of the bone loss we see is in the lower back, the femoral neck, and the greater trochanter, which is in the hip. So we focus on hip-dominant exercises — squats and deadlifting. Those are the main exercises we build the whole program around. They're also multijoint, multiplanar movements. We wanted to move his joints in as many different planes as we could, so we threw him everything just in case, so he'd be prepared. Maybe Soyuz lands five hours away from where Mission Control thought you were going to land, and you have to get out of the capsule, and it's 30 below zero. You have to be ready for anything. An astronaut doesn't have to be a great athlete, but he has to be fit overall. You're better off being good at many things than really good at one.[28]

Kimbrough pointed out that there was one thing his crew needed to be prepared for in space, even before the return: spacewalking, a physically demanding task that requires astronauts to work outside, sometimes moving massive objects around, for anywhere from six to eight hours. Not all astronauts do such activities; they are only scheduled occasionally for when big upgrades or repairs are required. But it is yet one more motivation for staying fit in space aboard the treadmill, exercise bike or resistive device. "It's really challenging to move this mass — the big, white spacesuits we have that weigh about 300 pounds," Kimbrough said. "Being able to control one takes strength and technique. Every time we open and close our hands we're fighting the pressure of the spacesuit, so they get worn out. We really do everything with our hands during a spacewalk and very little with our feet."[29]

NASA separates the risks to long-term astronauts into five categories, nicknamed RIDGE: Space Radiation (R), Isolation and Confinement (I), Distance from Earth (D), Gravity Fields (G), and Hostile/Closed Environments (E). Radiation is not really something that can be addressed after the fact, but during the mission NASA advocates a combination of shielding, radiation monitoring and some operational procedures, such as instructing astronauts to go into more highly shielded spacecraft during solar flares. The agency is also continuing research into improving radiation monitoring and detection, including better detectors, better procedures for using vehicle stowage, and materials to reduce exposure such as wall linings with radiation-resistant material. NASA is also continuing research into radiation on the ground and in space.[30]

The I, isolation, is something that many seniors are familiar with. Millions of older people are confined to their houses or to long-term care facilities, and isolation became even more acute during the coronavirus pandemic, which forced quarantines for senior safety before vaccines were widely available. NASA's recommended countermeasures before and after flight are numerous in this regard, including using analog research facilities like Antarctica's, installing light-emitting diodes to promote activity, encouraging astronauts to have safe and secure venting points such as written journals or speaking with medical professionals, and in emerging research, using tools such as virtual reality to simulate the effects of going into a new environment on Earth.[31] We can also group the Distance from Earth (or D in RIDGE) in this isolation section in terms of applications to seniors, as most of the space research in this area is about finding ways of enabling people to reach out to friends, family and co-workers from a distance. This is why so many seniors, for example, bought iPads and learned how to communicate virtually during the pandemic.

Within the RIDGE framework, NASA's research on gravity (G) has perhaps the most application to seniors. NASA has so many countermeasures available in space and on Earth to deal with the effects of gravity and microgravity that it is difficult to summarize them quickly (and we have discussed some of them already), but a short list would include functional task testing, fine motor skills testing, examining how fluids in the body affect vision as they shift (and how well compression cuffs and lower body negative pressure devices counteract this), countering back pain by using spinal ultrasounds, monitoring muscle size and bone density using MRIs, and occasional fitness self-evaluations to see how the body is responding to training, spaceflight and recovery. Exercise is an essential before, during and after flight, and drug countermeasures have also been used, the agency noted.[32]

The last component of RIDGE is hostile/closed environments (E). Studies show that living conditions of space can affect the immune system, although crews don't tend to fall ill when they get back home. "Even though astronauts' acquired immunity is intact, more research is needed into whether spaceflight-induced altered immunity may lead to autoimmune issues, in which the immune system mistakenly attacks the healthy cells, organs, and tissues present in the body," NASA says. "Beyond the effects of the environment on the immune system, every inch and detail of living and working quarters must be carefully thought-out and designed. No one wants their house to be too hot, too cold, cramped, crowded, loud, or not well lit, and no one would enjoy working and living in such a habitat in space either."[33]

We have ample evidence showing how dramatically younger astronauts age in space; those changes are reversible postflight and these findings can be applied to improve the health and lives of elderly people on Earth. But spaceflying seniors remain a small and select group — as we've seen, only a handful of very old people have flown to space. The eldest so far — Wally Funk and William Shatner — experienced suborbital flights of only a few minutes long in 2021; both appeared to be energetic in media interviews post-landing, although detailed medical results were not disclosed. The longest senior stint in space so far still belongs to NASA's John Glenn. According to the Smithsonian Institution's National Air and Space Museum, "There was understandable skepticism about the value of the biomedical research on a single person and suggestions that the experiments were simply a cover for giving the NASA champion a late-career victory lap." But what can't be disputed is that the 77-year-old had shown a healthy senior could thrive during an extended spaceflight — and that collecting data on people like him might be helpful for less fortunate seniors on Earth.

For spaceflying seniors like Funk, Shatner and Glenn, there is no turning back time; nothing experienced or learned in space can help them reverse the aging process, unlike younger astronauts whose aging-like symptoms are only temporary. (Indeed, although Glenn's good health continued for many years after his spaceflight, he died in 2016 at age 95.[34])

However, it is clear that studying astronauts young and old in space will be beneficial to older people on Earth. The big questions today are, which senior will be the next to participate in an extended orbital mission, and what new information may we collect from them? The key is to find someone willing to continue to participate in medical experiments — and to properly calibrate each result to match up with the more famous Glenn.

CHAPTER 12
Seeking Resistance

I n 2013, NASA astronaut Mike Hopkins was ready for his first flight, a six-month venture on the International Space Station that would test his body's ability to adapt to a floating environment and then get healthy on Earth all over again. When in training, it's common for astronauts to seek public engagement projects, especially to get children interested in science, technology, engineering and math (STEM). For Hopkins, figuring out what to focus on was a no-brainer. The astronaut has been active ever since he was a small child, and remains dedicated to exercise in his fifties.

"He grew up on a farm outside of Richland, Missouri, staying active by playing in the creeks and in the woods," NASA wrote in a mini-biography of Hopkins's exercise passion. "He started playing basketball in second grade but fell in love with football when he started playing in fourth grade. Hopkins continued playing football through college and became a captain for the Fighting Illini football team at the University of Illinois. Now his many hobbies keep him active. He spends time skiing, backpacking, running and participating in CrossFit. 'Staying fit and being physically active are important parts of my life,' said Hopkins."[1]

All astronauts need to be concerned about fitness to maintain their health in space, combat the symptoms of premature aging they experience in microgravity, and make sure they are in the best condition for their return to Earth's gravity. Resistance exercise is one of the most critical elements in making all of that happen, the strongest countermeasure to date for the deconditioning and atrophy that astronauts routinely experience in space. In a YouTube video, Hopkins showed off on Earth one of the more innovative devices used by crew on the International Space Station to keep fit — the Advanced Resistive Exercise Device (ARED), which we discussed in our brief tour of exercise devices earlier in the book. "It looks a little bit different than your squat rack that you might have in your local gym," Hopkins said. "But it has some of the same functions in the sense that we do squats on this. We can do deadlifts, bench press, shoulder press, things of that nature."[2]

Hopkins quickly did a tour of the device in front of the camera. "We actually have two canisters here that have vacuum in them," he said.[3] "And so we pull against that vacuum, and we can just dial in different loads with the handle . . . And we can change how much weight we're lifting." He then showed how each astronaut can adjust the bar to make sure it's at the right height, including rotating the handles individually. In fact, the settings the astronauts use on the ground version of ARED are identical to what is used in orbit, making that one less thing to remember the first time they're confronted with the exercise machine on the space station.

In a typical session, astronauts do about three sets of 10 exercises on ARED, but it can vary depending on whether they are in a heavier, medium or lighter day of exercise, explains a NASA Johnson Space Center video about the device.[4] ARED is compact and allows astronauts to continue to lift weights in space with devices such as vacuum cylinders and flywheel assemblies, which can operate independently of gravity and provide a workout.[5] "ARED accommodates a wide range of body types and sizes," the video added. "There is also a touch screen that makes it easier for an astronaut to follow a personalized, prescribed exercise plan . . . The crew performs their exercises using either a lift bar or a cable assembly. Resistive load can be adjusted between zero and 600-plus pounds for bar-related exercises, and up to 150 pounds for cable-related exercises."

Creating resistance in microgravity is a considerable challenge. ARED is an elaborate pulley system that is great for keeping astronaut muscles strong on the ISS for the return to Earth, but despite its compact design it still requires a lot of space. If you can imagine the challenge of fitting a set of barbells and a lifting bar inside a spacecraft, that approximates the difficulty of finding a spot to put ARED. Luckily, the ISS has multiple rooms available and ample space for ARED, but a typical spacecraft would not have such luxuries. What can astronauts do in the future for resistive exercises over long periods of time?

Often, when examining such questions, we can look to guidance from earlier space programs. But the years of sending astronauts to the Moon only offer an interim solution. Astronauts in those years used an Exer-Genie Exerciser, a technology that was basically off-the-shelf. The Smithsonian Institution's National Air and Space Museum (NASM) describes the device as a set of ropes.[6] It certainly was compact and was adequate for keeping the astronauts fit for their relatively short, two-week spaceflight.

"The Exer-Genie weighs less than two pounds and takes up very little room," NASM wrote. "A specially woven nylon rope goes around a metal shaft and pulling on the rope causes friction. In a spacecraft, astronauts hooked the device on the wall of the spacecraft. By pulling on the cord in a controlled way at varying speeds, astronauts could perform exercises closely related to isometrics (pushing or pulling against an immovable object) and isotonics (moving exercises such as calisthenics or weight lifting). The apparatus allows for more than 100 basic workouts, and has the added advantage of adjustable resistance, so each Apollo astronaut could set it to his own physical conditioning level."[7]

NASM noted the device did not produce any "significant results" in space, although it seems that many astronauts liked it. "After the Apollo 11 mission, Neil Armstrong stated they 'did a little bit of exercise almost every day. The Exer-Genie worked alright. It got a little hot and stored a lot of heat, but it was acceptable.' . . . Apollo 7 astronaut Donn Eisele, who had reported that his lower abdominal muscles ached from 'floating around in the seated position,' told mission control that he felt a lot better after using the Exer-Genie. His crewmates Walter Cunningham and Walter Schirra also reported the exerciser was a 'good deal.' Schirra even went so far as to say, 'One of the best "spacey" things we've had in years.'"[8] The device Apollo astronauts used was handy in the small space available in their spacecraft and appeared to be effective for the short missions, which typically lasted only about two weeks. ARED is clearly more effective for maintaining health and conditioning over longer missions, but the drawback is it takes more space and may not prove practical for use on future, more cramped spacecraft.

For future Moon missions, NASA plans to use the Resistive Overload Combined with Kinetic Yo-Yo (ROCKY) device, which was developed by Zin Technologies of Middleburg Heights, Ohio. At the time of this writing, NASA is planning a series of missions on the Orion spacecraft, called Artemis, that would put astronauts into a capsule-sized vehicle when they're not on the Gateway space station near the Moon or working on the surface. ROCKY is designed to take up only one cubic foot of room, and it essentially looks like a black box (the size of a shoebox) upon which astronauts can strap themselves, then use a rowing machine-like band for resistance.[9]

"Astronauts will be able to use the device like a rowing machine for aerobic activity and for strength training with loads of up to 400 pounds to perform exercises such as squats, deadlifts and heel raises, as well as upper body exercises like bicep curls and upright rows," NASA said in a description of ROCKY. "The device can be customized with specific workouts for individual astronauts. It will also incorporate the best features from a second device evaluated during the selection process called the Device for Aerobic and Resistive Training, or DART, developed by TDA Research in Denver, under NASA's Small Business Innovation Research Program, including a servo-motor programmed to deliver a load profile that feels very similar [to] free weights to the exercising astronaut's muscles."[10]

With a little more room available on Gateway and the lunar surface, NASA and its international partners are considering options for exercise in these roomier environments. One example is the forthcoming European Enhanced Exploration Exercise Device, being developed by the Danish Aerospace Company on behalf of the European Space Agency. According to the company's website description, "The E4D machine combines cycling, rowing, rope pulling and more than 29 other weight training exercises in a single machine. This provides more versatility and thus also enhances the effectiveness of the astronauts' daily workout in space."[11]

Naturally, the science fiction fans among this book's readership are probably screaming by now that there could be an obvious solution other than the focus on resistance exercise to maintain health and conditioning in space: creating a gravitational environment, or artificial gravity, away from the bounds of Earth. The idea is that a rotating space station would induce artificial gravity using centrifugal force, much like what we saw in the famous Stanley Kubrick film *2001: A Space Odyssey*, back in 1967. A 2021 NASA podcast with Bill Paloski, former director of the Human Research Program at Johnson Space Center, discussed how artificial gravity might be useful for long-duration space missions:[12] as the space station spins, the gentle rotation would create an artificial gravity environment, allowing the astronauts to enjoy the benefits of gravity and also reduce the medical problems caused by microgravity, which affect systems ranging from balance to bones.

The idea of spinning space stations was popularized in the 1960s by Wernher von Braun, a German scientist whose biography has caused some controversy over the years thanks to his role in developing rocket technology in Nazi Germany. But he was also noted as an incredible NASA engineer, leading the development of the Saturn V moon rocket, among other spacecraft. One of von Braun's ideas was to create a 250-foot-wide inflated "wheel" constructed of reinforced nylon.[13] Despite his work, however, the space stations of the modern era mostly look like Erector sets and don't spin. That's because constructing rotating artificial gravity facilities is a considerable engineering challenge.

As NASA's Paloski points out, designing the system is really the biggest hurdle. He describes a few concepts developed at NASA, including Kent Joosten's "stick-like vehicle" with a nuclear reactor on one side and the crew quarters at the other, and a baton-like nuclear thermal rocket vehicle made by Stan Borowski. According to Paloski, there were maintenance issues that "the engineering community didn't like. If you had to do an EVA [or extravehicular activity, such as a spacewalk], there were some questions about if you had to stop the rotation to do an EVA and then turn it back on." There also were concerns about complexities that come from rotating an entire vehicle, which requires more moving parts. Moreover, medically speaking, even a small centrifuge produces at least minor confusion for the human body. Although, as Paloski says, a typical human will adjust over time, and you might be able to train a person to expect the variable G-loads within the centrifuge, he adds that varying G-loads could also create some issues in moving the astronaut around the space station.

NASA continues to explore other, more compact solutions — including the Human Powered Centrifuge and the more recent Human Performance Centrifuge — that would allow astronauts not only to exercise but also to experience G-forces similar to Earth for at least an hour or two (during the exercise period). It may be that such countermeasures could help astronauts adapt more readily to Earth gravity after their return, but more research is necessary and, naturally, NASA will want to test the idea on numerous people before rolling it out to spaceflyers.

In the meantime, resistance exercise remains the best countermeasure to the effects of microgravity. Recent studies suggest that coupling resistance and aerobic training may be an even better solution to help astronauts stay healthy in space. There is a study underway called SPRINT that is looking to reduce both the amount of exercise time and exercise machine volume in space, to take advantage of the smaller spaces that the Lunar Gateway or Orion spacecraft may provide moonflyers. The study aims to reduce issues with muscle, bone and cardiovascular function by including "an increase in the intensity and a reduction in the volume of resistance exercises . . . [with] very short, but high intensity interval-type aerobic exercises, and start the exercise countermeasures as early as possible in the flight, preferably in the first week."[14]

In 2012, NASA Lead Exercise Physiology Scientist Lori Ploutz-Snyder recorded a short YouTube video explaining the importance of the study, as the work was just beginning in space.[15] "We include aerobic intervals, and that's how the study got its name, they sprint. So they do some sprinting on the cycle and the bicycle, where they go at high intensity for short bouts and then have a rest period. And we also have resistance exercise on the ARED that they perform three days a week . . . this experiment is evaluating a higher intensity exercise prescription that can be done for shorter amounts of time and less frequently," she said.

Part of what made the study possible in space was that a new ultrasound device had just been shipped to the orbiting facility, she added, noting that only one person had participated in the study at the time the video was recorded — but that it had gone very well. "We're using the new ultrasound system to make measurements of the leg muscle size, and the crew members scan themselves and make measurements of their own muscle size. And we compare that with the loads they're doing on ARED to determine how well they're doing for the resistance exercise, and we can adjust the prescription accordingly . . . Both from the implementation standpoint and from the performance standpoint, our first subject was a great success."

A 2020 *Nature* study based on this investigation on the ISS notes that the SPRINT regimen takes its inspiration from high-intensity/low-volume training (HIT), which "has been extensively documented in populations ranging from elite athletes to clinical patients. In addition to the time savings of shorter exercise sessions, there is evidence to suggest that HIT may elicit superior physiologic adaptations [better conditioning] compared to traditional lower intensity/higher volume training."[16]

The applications from learning about resistance training in space could help all of us in our quest to stay fit. During the COVID-19 pandemic, billions of people suddenly found themselves working from home and in many cases, reducing their activities outdoors or in exercise facilities to contain the spread of the coronavirus. The problem was especially acute for those families who live in small quarters, such as an apartment, with no room or perhaps the means for a treadmill, a set of large weights or any other bulky exercise equipment. Venerable publications like *the New York Times*, in February 2021,[17] were recommending compact alternatives to gym membership, such as resistance bands, yoga, foam rollers and ultra-compact exercise bikes.

Even before the pandemic, some of NASA's advances in resistance training methods and equipment were being applied in the public realm. "Inventor Paul Francis worked with NASA's Johnson Space Center in Houston to perfect SpiraFlex, springs that simulate the tension of lifting a weight without needing a physically heavy object," the agency wrote in 2020 as part of a larger retrospective[18] on how space station research was meant to benefit earthlings, too. Initially, Francis had decided to use metal springs, which would be resistant and hold their shape even when pulled. "But those failed faster than he wanted — after just around 10,000 cycles," NASA wrote. "Undeterred, he sought out an alternative material and came across an elastomer compound, which he enhanced and fashioned into spiral-shaped torsional springs of various dimensions."[19] Francis's collaboration with NASA would end up making important contributions to the health of humans on Earth as well as in space. He went on to license his SpiraFlex technology to Nautilus Inc., which was instrumental in developing the popular Bowflex Revolution home gym; and, NASA pointed out, "The springs soon became the basis of an exercise machine that was used on the space station for over a decade."

Francis's work on SpiraFlex eventually ended up being incorporated in iRED, which "used stacks of the elastomer spring disks, now known as FlexPacks, arranged in two cylinders that produced up to 300 pounds of resistance," NASA said. "A 16-week ground test showed exercising with the iRED produced the same results as using free weights."

While iRED was eventually replaced with the more robust ARED, Francis remained busy on Earth and patented his ideas for more consumer products. A newer company called OYO Fitness, for example, is trying to make resistive technology even more user-friendly for small spaces, with devices expected to weigh only two pounds but to include dozens of possible exercises. For example, the OYO Personal Gym uses the same SpiraFlex strength-training fitness technology in a device that looks much like a tiny steering wheel. It includes two 10-pound and one five-pound FlexPack of weights, ankle attachments and a door attachment to make the most of a single room.[20]

Treadmills are another exercise device that has been adapted for use in space. Originally intended for cardiovascular conditioning, using a treadmill may also help astronauts achieve "stable vision," also known as gaze stability, while on the device. Often taken for granted, stable vision helps us avoid obstacles and adjust our gait for the terrain in front of us, whether it's flat or hilly. It's also one of a number of factors that allow astronauts to readapt to walking after returning from a long-duration spaceflight. No matter how much you try to strap somebody down so they stay on the moving belt in microgravity, there still is an inevitable bit of up and down floating that occurs in between steps when the subject is exercising, especially at high speeds. NASA is now running more studies to understand gaze stability to assist astronauts, both in space and in the first wobbly hours after they return to Earth.

The ability to walk smoothly or dance gracefully relies on a complex interaction between what we see with our eyes, information we receive from the balance apparatus in our inner ear and feedback from our moving limbs. Try shaking your head back and forth, then up and down while you're reading this page and you'll notice that the words remain in focus. You've just experienced the benefits of gaze stabilization.

Stable vision is important for all creatures in motion, because it safely enables navigating complex environments and helps detect threats at a distance. You may have wondered why pigeons move their heads back and forth when they walk? When they take a step forward their head stays in place keeping the world in focus. With the next step the pigeon's head needs to catch up with its body and it quickly jerks its head forward to minimize the accompanying blurry vision.[21] Walking for a pigeon provides a sequence of in focus and blurry perspectives of its surroundings.

Gaze instability can be particularly problematic for humans. Mary Wisniewski described the challenge of trying to move to Michael Schubert, a physiologist and physical therapist at the Johns Hopkins School of Medicine. "My vision continued to get worse, and my world bounced and vibrated," she said.[22] She was suffering from damage to the part of the ear that detects head motion following previous surgery to remove a tumor growing on her acoustic nerve. Schubert was able to retrain her vestibular system to provide gaze greater stability.

For returning astronauts it takes time to build their terrestrial coordination. "Newly returned astronauts have shown instability in walking and in making jump landings and have experienced oscillopsia, [or] apparent bouncing of viewed objects," NASA said in a 2020 study about gaze stability called "The Effects of Prolonged Space Flight on Head and Gaze Stability During Locomotion," run by Johnson Space Center's Jacob Bloomberg. "The mechanisms of these abnormalities are not known, although changes in muscle tone, muscle activation patterns and vestibular function could contribute. Instability of gaze could result from changes in the coordination of head and body movements."[23]

This study describes various protocols to better understand the phenomena, including observing astronauts on a treadmill pre- and postflight, walking a path with or without visual and auditory cues, and making jump landings from a platform 30 centimeters (one foot) high. The overall goal, besides tiring out returning spaceflyers with odd wired-up helmets and epic jumps, was to show how spaceflight alters our sense of gaze. The treadmill results are the most relevant to our discussion of walking in space versus walking on Earth. "In the preflight treadmill tests of head-trunk coordination, pitch head movements were smoothly coordinated to oppose the vertical movements of the trunk, thereby helping to maintain a stable gaze," NASA wrote. But they saw the coherence decreasing "significantly" after a spaceflight, particularly when the astronauts were looking at a visual target farther away. They said that in space, the treadmill may have induced a regularity because treadmills, by their nature, demand a rigid pace and strict gait.

The hope is these various studies on resistance, exercise and conditioning will help inform NASA's preparations for landing people on the Moon later in the 2020s, and allow them to work in space for long periods with a minimum of deconditioning. It will be interesting, for example, to see how a treadmill will perform on the Moon, whose gravity is just one-sixth that of Earth. Sadly, the person who first developed a treadmill for the space shuttle will not be there to see it, however. Former shuttle astronaut Bill Thornton — who flew on STS-8 in 1983 and STS-51B in 1985 — died in early 2021 at age 91. His work on this key exercise device garnered him 60 patents that were associated with it. "I still am the only person that has ever run around the world on the treadmill that he designed," Thornton joked in a 2010 interview with WRAL-TV in Raleigh, North Carolina.[24] Ad astra, Bill.

CHAPTER 13
Bare Bones

"**U**sually, the biggest demon is not out there, it's what is inside your head. That was one of the most profound lessons I've learned in my life,"[1] Rick Hansen declared in a 2010 interview. The world renowned "Man in Motion" has dedicated his life to raising awareness of the importance of spinal cord injury (SCI) research. A 1973 truck accident in which he was a passenger threw him from the vehicle and left him paraplegic at age 15. Twelve years later he would embark on a two-year world tour that changed the perspective of millions of onlookers as he wheeled more than 24,000 miles through 34 countries, raising more than $26 million for spinal cord research and awareness.

The research is critical to developing better treatments for spinal cord injuries, their long-term effects and perhaps, one day, a cure. It can also help astronauts participating in long-duration spaceflights, working on the space station and in future going back to the Moon and on to Mars. The reduced mobility of individuals with SCI results in bone loss like that described in astronauts, and space medicine doctors hope that perhaps the solution to one condition may help the other.[2]

Perhaps the desiccated skeletons associated with Halloween cause most to think of bone as a "dead" tissue, a static frame that provides attachments for the muscles to enable us to move our limbs. In fact, bone is a living, dynamic tissue, continuously remodelling itself in response to the stresses and loads imposed on it. As we get older these processes change. Aging is associated with bone loss, also known as osteoporosis — literally meaning porous bones. Understanding the factors that cause it and developing preventive strategies and ways to treat it are of tremendous interest to doctors, researchers and those affected by it.

One way to think of bone is like a bank account.[3] New bone can be created — deposited into the account by osteoblasts, bone-building cells that compensate for withdrawals by bone-resorbing osteoclasts. In addition to aging, bone loss can be caused by a host of factors including poor diet, deficiencies of calcium and vitamin D, alcohol and tobacco use as well as immobilization and lack of exercise.[4]

Bone loss involves both a decrease in bone mass and a disruption of the bone's micro-architecture. Up to age 25 to 30, bone turnover is generally in balance between the cells actively building and resorbing bone. Between ages 30 and 50, bone density tends to stay stable. But after 50 there is a shift in this balance, favoring bone resorption over bone formation.[5] Over time, that can result in a deficiency of bone mass and an increased risk of fractures in the weakened bone. Keeping bones strong when we're younger is a good strategy to help reduce the risk of fractures in our aging frame as we get older.

Unlike muscles, which feel sore after a workout and show the conditioning effect of lifting weights as we "bulk up" and increase muscle tone, we can't feel or see the effects of exercise on our bones. The best way to assess age-related bone strength is to measure its density with an easy DEXA scan that can give accurate information on bone strength and risk of fractures.

Breaking one or more bones over a lifetime is not uncommon and a single fracture does not indicate a problem with bone strength. Rambunctious children can break a bone in the course of play, and many regular athletes and those engaged in extreme sports have had multiple fractures throughout their career, generally healing without problems. There are, however, risks associated with broken bones causing damage to surrounding tissues, which is particularly true with broken vertebrae in the back or fractures of the skull. The huge forces that cause the fractures can also damage the vulnerable underlying nerve tissue, or the fractures can become displaced, which can crush or shear the brain or spinal cord, resulting in devastating head injuries or paralysis.

If movement and exercise play an important role in maintaining bone density, what happens to the bones in the legs of paraplegic patients who are no longer able to move them? The lack of mobility results in leg bone loss at a rate of 2 to 4 percent per month, with bone density stabilizing within one to two years after injury.[6] This dramatic decrease in bone density significantly increases the risk of leg fractures, from 1 to 6 percent in individuals without SCI to between 15 and 30 percent in individuals with SCI.[7] Living with an increased risk of breaking bones is one of the many challenges paraplegic and quadriplegic individuals must face. They are not alone — many people with other conditions have brittle bones that easily break.

Wearing a T-shirt with the saying "Fragile Bones — Remarkable Spirit," Antonella Verderosa spoke with Long Island TV reporter Waldo Cabrera about the challenges of living with an inherited condition called osteogenesis imperfecta. "I've probably broken well over 100 bones, and it could be simply from turning in your sleep, or just simply sneezing. One time I fell off the couch and broke my arm and my leg. I kind of just live my life, and when a bone just happens to break then that happens."[8] Taniya Faulk has the same condition and said, "Nobody in my family has broken a bone, I've broken 92. There's 206 [bones in the body] and my aim is to break all 206 . . . Ironically, I've never broken fingers or toes. I just broke my wrist putting on a shirt. I do disability awareness, bullying, things like that, and sort of self-esteem workshops, and I just educate kids about different types of disabilities. I tell them my story, which tends to kind of get into a little motivation. But I also talk to them about, you know, it's okay to be different . . ."

Known as brittle bone disease, osteogenesis imperfecta is an inherited bone disorder that results in extremely fragile bones. It has been called the disease of the osteoblast.[9] "The osteoblast produces an abnormal matrix that does not respond to mechanical loads. In compensation, the osteoblast population increases, and osteoclast activity is raised, leading to a high bone turnover rate," reported Dr. Francis Glorieux in the medical journal *The Lancet*. He also found that the "increased rate of bone turnover is compounded by secondary bone loss induced by immobilization caused by lower limb fractures."

There have been many attempts to find treatments, but Glorieux said that "attempts to improve bone mass and structure with calcitonin, cortisone, growth hormone, parathyroid hormone, thyroxin, vitamins A, C, and D, and minerals (aluminum, calcium, fluoride, magnesium phosphate and strontium) have not been encouraging." But there may be hope with medications called bisphosphonates, which "decrease bone loss and slow bone turnover . . . bone mineral density and physical activity greatly increased, [and the] fracture rate decreased."

Commonly used to treat osteoporosis, bisphosphonates are a family of drugs that bind to the surface of bones and slow down the bone-resorbing action of osteoclasts.[10] They have been shown to increase bone density and reduce the risk of fractures in the elderly, and have also been effective in treating bone loss after spinal cord injury.[11] Perhaps these medications could help astronauts who spend months in space to maintain their bone density in a microgravity environment.

Long-duration missions on the International Space Station have confirmed rates of bone loss ranging from 1 to 2 percent per month. Most of the bone loss takes place in the lower back and lower extremities, while the upper extremities are not affected.[12] Perhaps this is not a surprise, as astronauts move around in space using their hands and arms to travel from handrail to handrail. With reduced use in the absence of gravity, the muscles in the legs and the postural muscles in the back become weaker and the bones lose their density.

What about future missions back to the Moon or a six-month voyage to get to Mars? What would happen to bone density? Would astronauts be at risk of breaking a bone if they fell getting off their spacecraft after it landed on the surface? Based on the observations of bone density in individuals with SCI, it is likely that the changes in the bone density of astronauts will plateau over time. It is also likely that the partial gravity of the Moon and Mars might be sufficient to maintain bone density, and an exercise program might help regain bone density after landing, as is the case when returning to Earth. Regardless, the prevention of spaceflight-induced bone loss is clearly an important strategy for deep-space missions.

Prevention starts well before a spaceflight. Exercise is an important part of the preventive strategy to optimize bone density, and high-impact aerobic exercise is particularly helpful because walking, running and jumping increase both cardiovascular fitness and the skeletal loading that is associated with maintaining bone strength. By stressing the bones, strength training can increase bone density, so it's not surprising to see astronauts hitting the gym to get in shape for their mission.

Vibration has also been suggested as an option to prevent bone loss, both age-related and spaceflight-induced. The idea is compelling: simply stand on a vibrating platform to maintain muscle and bone strength. It is not clear how whole-body vibration increases bone density, but some studies have confirmed it can have benefits similar to the effect produced by jumping, bouncing or resistance exercises.[13] However, its effectiveness in preventing bone loss has been equivocal[14] in osteoporosis studies overall and further research has been recommended before using it as a preventive strategy for spaceflight.[15]

If mechanical loading from walking, running, weight lifting or perhaps vibration can increase bone density, maybe astronauts might wear a tight-fitting piece of stretchy clothing to replicate that effect. This idea came from anecdotal reports of Apollo astronauts who noted it was difficult fitting into their spacesuits because they had become taller in space.[16] The now tight-fitting spacesuit forced the crew members' bodies back to their earth-bound height. Could this simple type of loading be used to help maintain bone density?

Engineers at the European Space Agency have developed the Skinsuit, which aims to replicate the loading on the skeleton, prevent lengthening of the spine and perhaps help maintain the density of the vertebral bones in the back. Some astronauts have reported back pain in space, possibly related to the stretching associated with becoming taller in space, which might also be controlled wearing the individually tailored Skinsuit.[17]

The idea of developing a garment to reduce the long-term effects of spaceflight is not new. The Russian space program developed dynamic compression clothing for cosmonauts to exert a deep compression force on the body that produced axial loads up to 40 kilograms (88 pounds) and provided resistance to movement. The Pingvin, or Penguin, suit has been used since the 1970s, but so far the effectiveness of the suit in preserving bone mass has not been quantified.[18] Nonetheless, scientists continue to pursue the idea and Massachusetts Institute of Technology researchers James Waldie and Dava Newman have developed a Gravity Loading Countermeasure

Skinsuit whose loading on the body mimics standing.[19] Perhaps one day these suits will be part of the preventive strategies used to maintain bone density in space.

Preflight diet can also play a role, and the best diet for bone health is a low-salt, high-calcium diet where protein comes from vegetable sources such as beans or quinoa.[20] Maintaining a balanced diet and a resistive exercise program in space are two of the primary strategies to minimize the bone loss experienced by astronauts.

There is no device that can be used to directly measure bone density in space,[21] but it is possible to get insight into the changes that are taking place by looking at fluctuations in urinary calcium levels.[22] Detecting urinary calcium is as easy as using a dipstick and has been used to measure the rate of bone loss in bed-rest studies that are designed to mimic the immobilization experienced in space. Developing new ways to monitor bone density for missions to the Moon and Mars will help determine if preventive strategies are working during both the microgravity phase of getting to the destination and the partial gravity phase of living there.

Aside from following prevention strategies to avoid spaceflight-induced bone loss, treatment is also an option; various approaches have been proposed, including the use of drugs to reduce bone resorption or increase bone formation.[23] Reducing bone resorption with bisphosphonates has been studied in combination with exercise as a countermeasure to prevent or slow the rate of bone loss in space. Scientists from the Japanese Space Agency JAXA have reported, "Bone measurements and comparisons of early model resistance exercise device[s] or the advanced resistance exercise device (ARED) show that ARED and bisphosphonates together improve essentially all measures of bone physiology during spaceflight."[24]

Another drug, potassium citrate, has been used to reduce the risk of kidney stones in space and in terrestrial healthcare. There is evidence that it is useful in preventing bone loss and it seems like a reasonable drug to help with the risks associated with bone loss in space.[25]

Are there any alternatives to medication? Artificial gravity was mentioned previously, and while it may be conceptually compelling, the practical realities of rotating a large space station are significant. So far, no space agency has felt this is a viable solution. Perhaps the challenge is simply an issue of scale. Rather than rotating an entire space station, would it be possible to create artificial gravity somewhere inside it? Some have suggested boots with a magnetic sole contacting a metal floor; while this might help astronauts walk instead of floating from handrail to handrail, they would not load the body the same way gravity does on Earth.

NASA's Ames Research Center pursued a different concept and has developed a Human Performance Centrifuge (HPC) capable of producing a gravitational gradient with forces up to five times that of the Earth's gravity at a subject's feet and one G at their head as they cycle around a circular track. Testing on Earth is done horizontally with the circular track on the floor.[26] In space, the six-and-a-half-foot track could be oriented at the end of a module, a seemingly great idea with perhaps a few challenges. The forces associated with the rotating HPC would be transferred to the space station structure, possibly interfering with the microgravity environment critical to some of the onboard science experiments.

The vibration caused by the use of the current space station exercise equipment has been isolated to prevent adverse effects on sensitive experiments when the crew is working out, but the effects of the HPC present a greater challenge. While it may be possible to find a similar engineering solution for the HPC, scientists need more information on how much intermittent exposure to artificial gravity is needed to prevent muscle weakness and bone loss. With further research and development, it may become an important component of the space gyms of the future.

After years of research, international partner space agencies have developed a reasonable set of countermeasures for middle-aged astronauts, but what might happen if people in their seventies and eighties were to fly in space? The question is becoming more important in the new era of commercial space travel and was highlighted by Wally Funk's July 2021 suborbital flight on the Blue Origin New Shepard capsule at age 82. Funk was arguably long overdue for a spaceflight. The pilot, flight instructor and former FAA inspector had gone through testing in the early 1960s to become a participant in the Woman in Space program with 12 other women, who became known as the Mercury 13.[27] Perhaps for political reasons more than operational concerns, none of these incredibly talented women flew in space at the time despite the Russian space program flying Valentina Tereshkova in 1963, earning her the distinction of becoming the first female in space. She eventually orbited Earth 48 times.

Shortly after Funk's flight, William Shatner captured global attention flying with Blue Origin at age 90. Both were short suborbital flights reminiscent of Alan Shepard's Mercury flight in May 1961 and although they were exciting to watch, they did not last long enough for changes in muscle strength or bone density to occur.

That was not the case for the former "oldest person in space." Senator John Glenn flew his second spaceflight in 1998 aboard the space shuttle at age 77. In a 2006 interview, Glenn commented, "I had served on a Senate Special Committee on Aging and had seen some parallels between the human aging process and some of the symptoms that occur with astronauts who spend a significant amount of time in the zero gravity of space . . . A good example is osteoporosis. I met with NASA to talk about the idea of studying these two related areas by sending an older person into space . . . This research would not only help astronauts be able to take longer space flights but would help with some of the frailties that we have in old age."[28]

The mission was a tremendous success, captivating the imagination of young and old alike. The research highlights focused on a slate of more than 80 experiments: "Medical research during the mission included a battery of tests on Payload Specialist Glenn and Mission Specialist Pedro Duque to further research how the absence of gravity affects balance and perception, immune system response, bone and muscle density, metabolism and blood flow, and sleep."[29]

There are many parallels between the deterioration astronauts experience in cardiovascular conditioning, muscle strength and bone density, and the changes seniors experience with aging. There has always been a "yin-yang" type of relationship between studies on Earth and those in space, with terrestrial research often helping humans travel farther and stay longer in space, while space research has helped transform the way we live and age on Earth. An exciting new area of research into the apparent accelerated aging associated with long-duration space travel, its underlying mechanisms and possible preventive strategies could help both astronauts and seniors improve their resilience and decrease the risk of disease.[30]

These studies assess senescence — the process of growing old — which entails a gradual deterioration of the functional characteristics of cells as they age. Senescent cells are unique in that they don't die; they eventually stop multiplying and remain in the body, continuing to release chemicals that can trigger inflammation and damage neighboring cells.[31] Researchers are now studying a class of drugs called senolytics that can selectively cause death in senescent cells and that may one day help seniors stay healthy as they age, while possibly preventing the deconditioning experienced by astronauts in long-duration spaceflight as they reach for the stars to pursue their dreams.

CHAPTER 14

Where's Up?

E gress. In space it is typically associated with opening the airlock hatch to exit the International Space Station and begin a spacewalk. Floating in the void 250 miles above Earth is, to say the least, compelling; some would say dizzying or overwhelming. Most astronauts will say, "It gets your attention" — an accurate but understated description of the visceral fear that some experience the first time they go out the hatch. Fear is an innate response shared across many species as a mechanism to protect against injury or death. There are obvious risks with exiting a relatively safe spacecraft to explore the vacuum of space, where temperatures can range from minus 250 degrees Fahrenheit to plus 250 degrees Fahrenheit,[1] so it is not surprising that internal alarm bells start ringing for many astronauts facing the final frontier.

Humans are capable of rational thought, which can overcome innate fear. Understanding where risks lie and how to control those risks helps prepare for fear-provoking activities. Whether it is skydiving or spacewalking, training plays a role; preparation can enhance performance. But as the saying goes, the plan may change the moment the hatch is opened, and experienced space-walkers understand the need to expect the unexpected. Despite rigorous training and abundant experience, it is not unusual to feel a little trepidation when leaving the safety of the mother ship. Fortunately, focus fights fear and spacewalking is a task that requires complete focus for success.

The airlock on the International Space Station is relatively small. It is designed that way for a good reason: every time a spacewalk takes place, the volume of air in the airlock is lost to space and must be replenished by the onboard life support systems. With limited supplies, efficiency is critical. With their spacewalking tools, the large backpack-like spacesuit life support systems and other equipment, only two astronauts can fit in the airlock at a time. They are positioned head to toe, for fit and function, as one of the astronauts must operate controls to depressurize the airlock at the beginning of the spacewalk and repressurize it on re-entering, while the other must open and close the airlock hatch located at the other end. This means that the first person out the hatch goes headfirst, while the second goes out feetfirst. For some, that makes a world of difference.

We are used to living and working under the influence of Earth's gravity. It very clearly gives us a clue that down is in the direction of the floor. That's the direction in which objects fall. Our language and life experiences are based on that ever-present force. Up is the opposite of down; it's pretty easy to figure out and all of us can readily identify one from the other. But it's a little different without gravitational cues to help define up from down, and falling doesn't have much meaning in a world without gravity. Life in space can be very disorienting.

Most astronauts somewhat revert to using their own bodies to orient themselves — their head is perceived to be up, and their feet are down. This works well until it is time to interact with the numerous switches, controls and interfaces on the various racks of space station equipment. The switches are labelled for good reason and the labels should always be read, not only so their meaning is clearly understood but to avoid potential catastrophe. In this situation, astronauts orient themselves so their body axis is aligned with the vertical direction of the text. Without gravity, the equipment interface implies the appropriate orientation to operate the equipment.

What happens when astronauts are not working with a rack of equipment? Perhaps two astronauts are floating in a module chatting with each other; how do they orient themselves during their conversation? Without gravity there are still cues to help with orientation. It seems a little strange to be talking with someone who is upside down — their feet are facing your head and vice versa. The more sociable solution is to orient in the same direction so that your head and feet are aligned with the other person's. This is particularly true when eating a meal in space; it is somewhat disconcerting to be swallowing your food while floating around chatting with someone's feet. Who determines which direction all will share in aligning themselves for a sociable meal? Does some alpha astronaut dominate the group? Perhaps everyone watches surreptitiously to see which way the commander is oriented? Fortunately for cosmonauts, in the ISS's Russian segment there's no need to worry as the crew often eat floating around a table.

But in the airlock the two astronauts working head to toe will exit in different orientations. The lead spacewalker opens the hatch and exits headfirst. Most are amazed by the spectacular panoramic view of Earth above them; the stunning image is one they will never forget. After double-checking the tether, there's time for a couple of minutes enjoying the view while waiting for their colleague to exit.

The experience is slightly different for the second astronaut, who, exiting feetfirst, perceives Earth to be below them. NASA astronaut Joe Acaba described the moment: "It does take a little bit of work while you're doing a spacewalk to realize, OK, it's OK. You are not going to fall."[2] He is not alone. NASA scientists have reported that "while space-walking, many astronauts report height vertigo — a sense of dizziness or spinning — that is often experienced by individuals on Earth when looking down from great heights. Some astronauts also experienced transient acrophobia — an overwhelming fear of falling toward Earth — which can be terrifying."[3]

It can be a little disconcerting to alternately feel that Earth is above or below you based on constantly changing body orientation during a space-walk, and it is easy to understand why some astronauts might firmly grasp the handrails as they move hand over hand around the outside of the space station. But experience helps control these illusions. It gets easier to enjoy the incredible view with greater spacewalking experience!

Illusions are described as "a misrepresentation of a sensory stimulus, an interpretation that contradicts objective reality as defined by general agreement."[4] Gravity — it's a law of physics and a universal force understood by scientists. Its ever-present nature provides a strong cue that helps us stay oriented with our surroundings. Aerobatic pilots can fly an impressive array of maneuvers and still know where down is, due to the force of gravity pulling them into or out of their seat.

Orientation, understanding where we are relative to our surroundings, is based on a combination of different sensations. Gravity helps us differentiate up from down with continuous feedback from feeling its effect when lying, sitting, standing or moving about. Visual stimuli are also important in providing cues about our surroundings. Lights hang from the ceiling, furniture rests on the floor, artwork is hung vertically or horizontally on walls so we can appreciate its beauty. But it is possible to become lost or disoriented without directions or other usual cues. Staying oriented is a good thing in space — no one wants to get confused about where they are during a spacewalk — and scientists have spent considerable effort studying how humans interpret different sensory inputs to remain oriented.

Canadian psychologist Dr. Ian Howard, who founded the Centre for Vision Research at Toronto's York University, was renowned for his studies of human visual perception and spatial orientation. One of his research protocols used a room that would rotate around the visual axis of the seated subject. The seat was located against the back wall of the room, and the remaining three walls, floor and ceiling were decorated with a number of objects: "The seat was on one wall of the room, giving the occupant a view of the rest of the room, consisting of conventional furniture including a coffee table with (at one stage) a *Time* magazine with Mikhail Gorbachev's photograph on the cover resting on it, a light fitting hanging from the ceiling, a window on the left wall giving a view of a photograph of a natural scene, and the closed entrance door on the right wall."[5] Howard enjoyed having visitors to his lab participate in the rotating room experience. His sense of fun and curiosity was infectious, and most visitors readily agreed to be strapped in to see what this experience was like.

Using a five-point restraint like that found in racing cars or aerobatic airplanes, the visitors were strapped in, the door to the room closed and the fun began. After a couple of minutes of casually sitting alone in the quaint surroundings, the room slowly started to roll around its axis, giving the seated visitor a sense that they themselves were moving. All the visual cues in the room created the sense of being in motion, rotating on the "magical chair." The view when the room was upside down relative to the observer was particularly perplexing — visually one felt like they were upside down, now on the ceiling of the room looking down at the floor. Yet intellectually they knew that all the gravitational cues normally associated with being inverted weren't present. Their hair wasn't falling from the top of their head, and they didn't feel the force of gravity against their shoulders as they would if suspended upside down in the harness. In fact, the gravitational cue of being seated remained constant. The compelling visual cues overrode the sensations caused by gravity; made them feel *they* were rotating, not the room; and convinced them they were upside down when they weren't!

The sense of self-motion the subjects described is called vection, an illusion that can be quite startling. It can occur in different directions, and while the subjects in Dr. Howard's experiment experienced rotational vection, most of us have experienced linear vection — perhaps without appreciating what was happening.

One of the most common examples of linear vection takes place in the train station. Passengers seated by the window will sometimes feel their train is going backward when in fact it is not moving at all. The limited view of the train next to them captures their attention, and when that train slowly starts to move forward, many think that the adjacent train is stationary and they are moving backward. "Wait, we're going the wrong way!" they're thinking. And just before they want to call out to the conductor, the adjacent train is going so fast the illusion disappears. They realize they're stationary and think, "Am I ever tired" or "It's been a tough day!"

A similar illusory sense of self-motion can occur when driving a car that has stopped at a red light on a hill. If the peripheral vision perceives the adjacent car inching forward in anticipation of the light turning green, it can give a very convincing sense of rolling backward oneself, often resulting in confusion, firmer application of the brakes or pulling on the parking brake. As in the train example, the sense is short-lived and can often be followed by an embarrassed smile to the passenger seated beside the driver. Clearly, despite gravitational cues, changing visual stimuli can create the illusion of self-motion — the sense of vection — and there's no better place to assess the role of gravity in vection than space!

Understanding the mysteries of the mind is one of the most complex challenges for scientists studying the brain. A 10-year commitment to neuroscience was made when President George H. W. Bush declared the 1990s the decade of the brain. "The human brain, a 3-pound mass of interwoven nerve cells that controls our activity, is one of the most magnificent — and mysterious — wonders of creation. The seat of human intelligence, interpreter of senses, and controller of movement, this incredible organ continues to intrigue scientists and layman alike."[6] NASA responded with a 1998 space shuttle mission dedicated to neuroscience called Neurolab. It was a unique chance to understand how the brain adapts to being in space and a perfect opportunity to study vection.

A team of scientists was assembled, led by Chuck Oman from the Massachusetts Institute of Technology and including Ian Howard and other scientists from MIT and York University. The shuttle crew used a virtual reality headset called the VEG (virtual environment generator) to present various graphic images and videos to the crew. Some of the still images showed a detailed drawing of an empty space station module, while others included a person in different orientations. The Neurolab crew was asked to identify the direction they felt was up. The graphic image tests found that crewmembers differed on whether their subjective vertical was influenced by the orientation of objects they were looking at, or whether subjective down was in the direction of their own feet.[7] Some of the astronauts felt that the position of their head indicated which direction was up, whereas others were influenced by the orientation of the subject in the image. However, the tilted and tumbling video scenes revealed that most crewmembers felt the floor was the surface beneath their feet. The varied responses suggest there may be several ways to perceptually adapt to weightlessness and many ways to define where up is in the absence of gravity. Some astronauts have reported that the labels on the racks of hardware are an overriding cue and they rotate their bodies, aligning vertically to be able to read the label right side up.

The vection studies demonstrated an earlier onset and increase in the sense of self-motion when compared to the data collected on Earth. Personal anecdotes described the sensation a little more dramatically. "It felt like the movie *Ghostbusters* where you see the ghost flying down the corridor, except in this case the corridor appeared infinite."[8] Not surprisingly, the absence of gravitational cues changed the way the crew interpreted visual cues about their environment. The onset and magnitude of the sense of self-motion was faster and greater in the absence of gravity.

This observation underscores the possibility of shifting the priority of different senses to obtain a more accurate perception of the world around us. For the most part humans use their five primary senses, but information from at least five additional senses and, some researchers believe, up to 18 different sensory inputs could be used to accurately interpret our surroundings.[9] In addition to the five primary senses they use every day, most people would add to the list the ability to sense temperature, pain, balance, thirst and hunger. The ones we don't think about include proprioception; the subtle differences between different types of touch, including itching, light touch and pressure; and internal sensors that monitor muscle tension, stretch receptors, chemoreceptors and possibly some sense of detecting magnetic fields.

Our ability to correctly orient ourselves to the world around us is fundamental to living independently and can be critical to our survival. It is a skill that is shared among different species, whether it be whales and dolphins using echolocation to help them navigate, giraffes using visual cues to find the watering hole, or humans trying to find their parked car at the end of the day. For the most part we take it for granted until we become temporarily disoriented or, even worse, we misperceive visual cues while driving and end up in a car crash.

The unique aspects of human spaceflight cause astronauts to develop new strategies to help them remain oriented. It is an excellent opportunity for researchers to study how quickly those new strategies develop, which senses predominate and how accurately their perception reflects their surrounding environment. Teaching people how to use different senses to overcome the loss of a primary sense is a proven strategy that might also be effective in helping patients who have problems with stability, such as those with Parkinson's disease or muscular dystrophy.[10] Perhaps the data from space will one day help clinicians develop different approaches to assist patients with building new strategies that rely on different sensory information, helping them live their lives to the fullest.

CHAPTER 15

Tricorders and Holodecks

I n a virtual reality room nicknamed the "holodeck," the android robot Data and Captain Jean-Luc Picard practiced a performance from the William Shakespeare play *The Tempest* — an old favorite of both crew members. Unfortunately, the computer suddenly generated a locomotive and sent it their way, not only interrupting the rehearsal, but forcing the duo to leap to safety. Data later used his tricorder device to radio another crew member, telling the audience the obvious: the holodeck safety protocols weren't working and any injuries sustained in the virtual reality world could be lethal.

This 1994 episode of *Star Trek: The Next Generation*,[1] which appeared very futuristic in its day, previewed technologies that are now typical to the experience of many American children comfortable with technology and in households that can afford it. Virtual reality headsets simulate the experience of the holodeck to some degree without needing to wire an entire room on a starship to help astronauts escape the monotony of another day of deep-space operations (and without the danger of events in the virtual world breaking through into the real one).

Many technologies developed on Earth have been repurposed to make life in space easier and richer, and, vice versa, space and science fiction have inspired many technologies that improve life on Earth. Virtual reality and gaming are among them. The tricorder comes to us in the form of cell phones, which allow us to stay in touch with people all over the world and gain information through means as diverse as texting, surfing the web, making calls and watching video, depending on your bandwidth. Now that it's possible to gather health data such as heart rate, blood sugar levels and EKGs through many smartphones and even smart watches, the medical functions of the tricorder are already beginning to be realized, too. And with SpaceX and OneWeb deploying satellites to make cell access available in remote areas, the expectations for constant connectivity continue to rise. A cute *PC World* article from 2017, called "How to transform your smartphone into a real-world *Star Trek* tricorder," suggested adding apps focusing on photo recognition, satellite tracking, temperatures, and even motion detection to simulate the tricorder experience of seeking strange new lifeforms and performing medical experiments.[2] Later that year, the Qualcomm Tricorder X Prize competition awarded the top prize of $2.6 million to Final Frontier Medical Devices, a Pennsylvania-based team; news reports described how similar the winning team's device was to a *Star Trek* tricorder.[3] The company developing it today, Basil Leaf Technologies, affectionately calls the tricorder "DxtER." Here's how they describe it on their website: "At the heart of DxtER is a sophisticated diagnostic engine based on analysis of actual patient data and years of experience in clinical emergency medicine. We developed algorithms for diagnosing 34 health conditions, only half of which were leveraged in the competition. Some of these conditions include: diabetes, atrial fibrillation, chronic obstructive pulmonary disease, urinary tract infection, sleep apnea, leukocytosis, pertussis, stroke, tuberculosis, and pneumonia."[4]

We also can't forget the tablet, which *Star Trek: The Next Generation* showed constantly; think about Picard sitting casually in his ready room, Earl Grey tea in hand, browsing reports from Starfleet with what the 1990s thought was a far-future device. "One interesting characteristic of *Star Trek: The Next Generation* — one that separated it from the original series and most of the early films — was its widespread use of smooth, flat, touch-based control panels throughout the Enterprise-D [starship]," wrote the Ars Technica website. "This touch interface was also used for numerous portable devices known as PADDs, or Personal Access Display Devices. These mobile computing terminals bear a striking resemblance to Apple's iPad — a mobile computing device largely defined by its smooth, flat touchscreen interface."[5] Ironically, part of the reason the sleek device came about, the article added, was the TV show's relatively low budget, which didn't allow for the usual gauges and switches present in *Star Trek* feature films.

The *Star Trek* franchise reminds us that humans adapt to space, and to remote environments in general, through a blend of biology and technology. The 2020s offer an exciting opportunity to make it easier to simulate medical procedures or to work through tough technical tasks using artificial intelligence and augmented reality. NASA is among the many space organizations hopping on board this revolution, even adapting what works in gaming to help astronauts train for the rigors of spaceflight. Since the technology changes so quickly, here we will focus more on what opportunities these types of innovations bring to crew members, instead of discussing specific capabilities.

First, a primer on which gaming technologies can be repurposed for educational purposes and, beyond that, in training for spaceflight. No matter what platform video gamers use, there tend to be a few elements universal to good game design (and with one of the authors being a regular gamer, this description comes from experience). Any platform usually teaches gamers how to perform simple tasks through a set of training exercises, which can take anywhere from seconds to hours depending on how complex you want the experience to be. While we must remember that every game has new players who highly depend on the "onboarding" experience, there is a language to PlayStations, Xboxes and other console devices that is fairly universal.

"Modern games have many conventions to fall back on, and no longer have to explain to the player that 'the left joystick moves the character, and the right joystick controls the camera,'" wrote Avantis Video's Yahel Galili. "In fact, wasting the player's time and stopping the action to cover this basic info will likely break the player's immersion. For another example, games commonly indicate climbable walls with clear, colorful markings. These serve both as a convention that's instantly recognizable to experienced players; and to imply an affordance that even novice players will notice and attempt to interact with."[6]

Once the gamer achieves mastery of certain tasks, the game tries to simulate the experience of "flow," which has been studied and elegantly described by psychologist Mihaly Csikszentmihalyi: "There's this focus that, once it becomes intense, leads to a sense of ecstasy, a sense of clarity: you know exactly what you want to do from one moment to the other; you get immediate feedback. You know that what you need to do is possible to do, even though difficult, and sense of time disappears, you forget yourself, you feel part of something larger. And once the conditions are present, what you are doing becomes worth doing for its own sake."[7]

Note that the word "flow" isn't always agreed upon in the academic community. A 2018 article in the journal *Frontiers in Psychology* notes that many elements of flow can be substituted for immersion.[8] Quoting a 2004 study, the article says there are three grades of immersion that have been identified: engagement, engrossment and total immersion. Getting to that latter stage may not be achievable, depending on how the player is feeling and how well the game is designed. "Hence, the model of an average immersive experience in video game playing can be reduced to the engagement and engrossment levels, whose characteristics are not considerably divergent from flow," the study says.

In a tiny percentage of people, the immersive experience of video games can be detrimental. Many gamers report minor incidents of, say, going to bed half an hour or an hour late as you try to beat that tough "boss" to level up. The World Health Organization, however, noted a phenomenon known as "gaming disorder," defined as "impaired control over gaming, increasing priority given to gaming over other activities to the extent that gaming takes precedence over other interests and daily activities, and continuation or escalation of gaming despite the occurrence of negative consequences."[9] Before making your kid throw away the Nintendo, however, the WHO tells us that classifying it as an addiction should be done with care (and probably should involve a discussion with a neutral physician, for a start): "Studies suggest that gaming disorder affects only a small proportion of people who engage in digital- or video-gaming activities. However, people who partake in gaming should be alert to the amount of time they spend on gaming activities, particularly when it is to the exclusion of other daily activities, as well as to any changes in their physical or psychological health and social functioning that could be attributed to their pattern of gaming behavior."[10]

Whether you're talking about flow or immersion, the game reinforces the feeling of mastery for players by including such gameplay elements as overcoming difficult challenges or boss battles, earning trophies by achieving certain objectives, or unlocking previously inaccessible content after completing a difficult task — for instance, making more areas of a map available to players. In role-playing games, players gain skills that will make fighting and other tasks, such as unlocking chests, easier once they accumulate "skill points."

These principles of "gamification" can be adapted for numerous life situations, including motivating children to complete chores[11] and enhancing leadership development training in the workplace.[12] Gamification thus can be applied to education.[13] For example, the website Khan Academy encourages children from kindergarten to Grade 12 to continue learning mathematics, computer science, English and other common subjects through tools such as badges; beginners are hooked by winning the simple "meteorite" badges, and advanced learners can compete for ultimate mastery with "black hole" and "challenge" badges.[14]

To be clear, gamification doesn't necessarily mean a game-like experience, so leaning on this concept may not get your reluctant child learning math unless there is some other motivational reason aside from badges. Karl Kapp, an instructional technology professor at Bloomsburg University, distinguishes between structural gamification and content gamification. The structural form allows the learner to "go through the content and to engage them in the process of learning through rewards," such as receiving points simply for completing a module or assignment. Content gamification is more rich: "This is the application of game elements and game thinking to alter content to make it more game-like. For example, adding story elements to a compliance course or starting a course with a challenge instead of a list of objectives are both methods of content gamification. Adding these elements makes the content more game-like, but doesn't turn the content into a game. It simply provides context or activities which are used within games and adds them to the content being taught."[15]

Now let's apply the principles of gamification to spaceflight. But first, let's make something clear: astronauts, cosmonauts, taikonauts and other spaceflight participants are quite fortunate to receive the opportunity to fly into space and they are highly motivated to do whatever it takes to get there; they are hugely committed to the endeavor long before they ever get off the ground. Unlike reluctant or struggling students on Earth, they don't need elements of gamification to coax them through their training. And while numerous private companies are working to bring down the cost of astronaut seats, fewer than 600 people have made it to space as of 2021, largely under government programs in which the individual did not pay their way. Those that did fly by other means typically paid millions of dollars for the opportunity, or won a contest; that said, the barrier to entry may diminish as companies such as Axiom Space, Blue Origin and Virgin Galactic ramp up their nascent spaceflight plans. For now, getting to space is still very rare and the experience is confined to a select few who are incredibly focused on the objective.

Those who do make it to space often experience what space philosopher Frank White termed the "overview effect,"[16] in which the normal boundaries of human experience like culture, language and border fade away through the tremendous experience of seeing Earth from far above. There are therefore moments of transcendence in space, best represented by times such as looking out the window, participating in a spacewalk, showing off somersaults to elementary students on camera, or even going through the rituals of launch and landing.

Spaceflyers all over the world have reported similar feelings associated with the overview effect, with one example coming from billionaire space tourist Anousheh Ansari: "The actual experience exceeds all expectations and is something that's hard to put to words . . . It sort of reduces things to a size that you think everything is manageable. . . . All these things that may seem big and impossible . . . We can do this. Peace on Earth — No problem. It gives people that type of energy . . . that type of power, and I have experienced that."[17] The experience also has been reported by numerous professionals, such as former test pilot Ed Mitchell, part of the Apollo 14 crew and the sixth person to walk on the Moon: "There was a startling recognition that the nature of the universe was not as I had been taught . . . I not only saw the connectedness, I felt it. . . . I was overwhelmed with the sensation of physically and mentally extending out into the cosmos. I realized that this was a biological response of my brain attempting to reorganize and give meaning to information about the wonderful and awesome processes that I was privileged to view."[18] This powerful overview effect goes beyond the flow or total immersion experienced in gaming and is shared by only a few hundred humans. One of the hopes of space tourism is that more people will be able to experience such feelings, although of course true affordability is decades away at the least.

These are some of the *good* things an individual feels about spaceflight, in addition to feeling that you are contributing to and expanding on important research, to the benefit of individuals on Earth (such as seniors, or remote communities, both of which we discussed earlier). But space can also be a monotonous and difficult experience, if not treated with the right care. In practical terms, a typical spaceflight sees the person cooped up in relatively small quarters with a group of people they may not necessarily be friends with, although of course the typical astronaut is trained to be a good team player and can overcome disputes as long as everyone is committed to the principle.

Other aspects begin to come into play, too; long separation from friends and family can occasionally result in missing out on life's milestones, like a birth or death in the family. NASA has a policy of communicating bad news quickly to astronauts, such as in 2007 when Dan Tani's mother died in a car crash while he was in orbit.[19] And NASA astronaut Frank Culbertson had the unfortunate experience of being in orbit when the 9/11 attacks on New York City occurred — and being the only American on the space station at the time,[20] meaning that (as sympathetic as his crew was) much of his support came from the ground.

We also must not forget that NASA has numerous psychological safeguards for its astronauts before, during and after missions, along with procedures for the astronauts and their families to work through in case of tragedy in orbit. And, certainly, these kinds of protocols are not unique to NASA. Former Soviet cosmonaut Georgy Grechko once told the Associated Press that Soviet authorities decided to withhold the news of his father's death until he finished a 96-day mission, which meant Grechko didn't find out about it until he landed. That said, the new Russian Federation did inform Vladimir Dezhurov immediately after his mother died in 1995, while he was on board space station Mir. (Crewmate Norman Thagard said that as Dezhurov and his crewmates struggled to absorb the sad news, NASA thought carefully about how to deal with tragedies in space, leading to some protocol changes.[21])

Beyond occasional major life events and interpersonal dynamics that can sometimes spice things up, a typical space day can feel overscheduled and, at the same time, somewhat monotonous. Think about having an entire Saturday taken up with meal prepping, cleaning toilets, exercising and doing a whole lot of mundane things necessary to keep you and your body functioning well. Then transport that experience to space, where you not only have to do all of these things regularly, but most of your communications are long-distance and your crew members are so similarly busy that the only times you can really catch up are meal times, if you're lucky. Highly scheduled days, monotony and frustration can easily catch up to you in space, if you're not careful. NASA has numerous safeguards against that, fortunately, and gaming can be one of them — along with other leisure time activities like playing an instrument, sewing, taking photographs or calling friends and family, which is just a selection of what astronauts have done on the International Space Station to pass the time and relieve the boredom.

To be sure, space can also induce the opposite effect on a crew under the right circumstances, which is the wonderful feeling of a group of people coming together to work at a high level of performance and adapt to the environment. "Surprisingly, some studies suggest space travel can actually be good for an astronaut's mental well-being," the Massachusetts Institute of Technology said in a 2019 blog post,[22] adding: "In a 2007 study, psychiatric researchers Dr. Jennifer Ritsher, Dr. Nick Kanas, Dr. Eva Ilhe, and Stephanie Saylor conclude, based on a group of previously conducted surveys, that space missions provide an example of salutogenesis, a process in which people are positively impacted by having to adapt to a harsh and stressful environment. This same phenomenon happens to some individuals in similarly nigh-uninhabitable environments, such as researchers at polar stations or people who travel in submarines."

But for those astronauts who struggle, virtual reality could bring a fully immersive experience to crew in remote environments to help them feel like they're back on Earth. For example, Stardust Technologies' Project EDEN will use virtual reality, artificial intelligence and a "haptic feedback" suit that simulates wind, touch, rain and other sensations that come from being outside. "The company's CEO, Jason Michaud, said the project is intended to help astronauts with feelings of homesickness, loneliness, isolation, and stress. He compares it to feelings many people on Earth have been experiencing during the pandemic," CTV News reported in 2021.[23] While the technology is still under development, the company was targeting the International Space Station as the next logical step in seeing how well astronauts enjoy a convincing reminder of home.

Lest we sound like a holodeck is just an investment in astronaut mental health — which is, of course, important but not all there is to a spaceflight — there are training benefits to virtual reality as well. NASA is still exploring its potential, but we can point to a few ways in which virtual reality is helpful. One benefit is known as "just-in-time training,"[24] which can deliver new information to crews already in orbit and supplement the many hours of training that they received at home.

A couple of examples may serve. All astronauts are certified in spacewalking and comfortable dealing with a range of normal and emergency situations, but it may be that an ammonia leak occurs on the station and, given the urgency to fix it and the inherent danger in the task (ammonia is highly toxic and can cause lung irritation), a virtual reality simulation would be helpful to make sure astronauts are as prepped as possible to deal with it swiftly. Another common problem is maintenance on the space station; considering just how many systems and subsystems are in operation, even 2.5 years of training is not enough to show astronauts all the things that can break. Just-in-time training can serve as an instant update to help astronauts jump into new tasks, extending what they have already learned to adapt to a fresh situation.

Project Sidekick is a virtual-reality collaboration between NASA and Microsoft that was running during the first one-year crew mission on the International Space Station in 2015–16. Sidekick was also tested underwater during the NASA Extreme Environment Mission Operations (NEEMO) 20 mission in 2015. NASA explained the project in a press release that year[25]:

> Sidekick has two modes of operation. The first is "Remote Expert Mode," which uses Skype, part of Microsoft, to allow a ground operator to see what a crew member sees, provide real-time guidance and draw annotations into the crew member's environment to coach him or her through a task. Until now, crew members have relied on written and voice instructions when performing complex repair tasks or experiments. The second mode is "Procedure Mode," which

augments stand-alone procedures with animated holographic illustrations displayed on top of the objects with which the crew is interacting. This capability could lessen the amount of training that future crews will require and could be an invaluable resource for missions deep into our solar system, where communication delays complicate difficult operations.

As NASA indicated, just-in-time training will be even more crucial during a long-duration mission, such as going to Mars. Astronauts may need reminders — after at least a few months of travel — to review and prepare for landing and setting themselves up on the surface safely, along with cautions and reminders to take care lest somebody slips or breaks a delicate hip due to a maladjustment from microgravity. This kind of training will be even more crucial because it's impossible for someone on the ground to talk you through procedures in real time. Crews on Mars will be on their own because of the distance, with a phone or radio signal taking 26 minutes, on average, to travel to the Red Planet and back.[26]

Another technology that may be useful to astronauts is "augmented reality." One of the most famous AR applications was a game called *Pokémon Go*, which encouraged you to chase a set of well-known creatures around the real world. As Vox explained in 2016[27]: "In simple terms, *Pokémon Go* is a game that uses your phone's GPS and clock to detect where and when you are in the game and make Pokémon 'appear' around you (on your phone screen) so you can go and catch them. As you move around, different and more types of Pokémon will appear depending on where you are and what time it is. The idea is to encourage you to travel around the real world to catch Pokémon in the game."

The game was not without controversy at its release, as it did not distinguish well between private property, public property and sensitive situations. One of the more infamous problems cropped up when Washington's Holocaust Museum begged people not to play the game in an institution dedicated to solemn remembrance of the victims of the Nazi regime in Germany and other countries.[28] Security and privacy observers also raised issues with the amount of information the game collected from users,[29] as well as the inherent problem of letting people — especially children, who the game specifically targets — roam around with most of their attention focused on their phones rather than, for example, traffic in the street. Yet augmented reality may prove useful for astronauts by providing a sort of "heads-up display" of their environment, similar to what pilots have in advanced aircraft or that robots had in the *Terminator* franchise.

NASA and the U.S. Navy are researching augmented reality in a range of operational environments. A 2017 article[30] highlighted a Diver Augmented Vision Device (DAVD) intended to be used on underwater missions, providing sonar information about the environment along with real-time information channels such as text messages, photographs and diagrams. The system was tested during NEEMO 23 and could also be ported to the challenging environment of the Moon's surface or the harsh terrain of Mars. "This capability is game changing for divers who usually work in zero visibility conditions — it essentially gives them sight again through real time data and sonar," Allie Williams, DAVD team lead engineer, said in the article. "Even in good visibility conditions, the DAVD system allows for hands free information and less mental strain of trying to remember topside instructions. The same benefits can be gained by astronauts as well — including better situational awareness, safety, and allowing them to be more effective in their missions." It is possible that this technology will help astronauts in orienting themselves in space as well as in maintaining their general health, but more study will be required.

Medical research is also advancing in terms of gaining full-body awareness through sensors either worn on the body or ingested, and some experiments have already taken place in orbit. Montreal-based Carré Technologies flew its Hexoskin, a prototype suit, to the International Space Station during Canadian astronaut David Saint-Jacques's 2018–19 mission. "This sensor system would continuously record, manage, and analyze crewmembers' physiological data such as vital signs, sleep quality, and activity levels. The data from the sensors would be transmitted wirelessly back to a software application that could be consulted in real time by the medical crew," the Canadian Space Agency wrote in a 2015 blog post.[31] "This new technology, currently being developed, has the potential to become an essential tool as part of an integrated medical system for human exploration missions. For example, in the event of an injury or illness, data from these wireless sensors, as well as other data collected during the mission, could assist the crew in establishing a quick diagnosis and a treatment plan."

A 2018 report by the CSA identified more than 2,000 companies in Canada alone that have the potential to offer medical technologies for space and remote environments, in areas as diverse as medical imaging, artificial intelligence–driven clinical decision support and remote monitoring. AI, in fact, has been portrayed in science fiction often; think about the role of C-3PO in interpreting data for Han Solo during a typical *Star Wars* sortie, or the adorable TARS and CASE robots of *Interstellar*. Newer advances in the technology suggest that before long, computers will be able to make decisions to assist with spaceflight, noted a 2021 article in Analytics Vidhya, the world's leading online data science community.[32] "Scientists are developing AI-based assistants to aid astronauts in their mission to Moon, Mars, and beyond. These assistants are designed to understand and predict the requirements of the crew and comprehend astronauts' emotions and their mental health and take necessary actions in the case of an emergency. Now how do they do that? The answer to this is sentiment analysis . . . [which] is a sub-field of NLP (Natural Language Processing) that tries to identify and extract opinions within a given text across blogs, reviews, social media, forums, news, etc."

These various developments in body sensors and AI begin to point to new trends in space exploration that will be better explored in the coming years. As computers become more adept in classifying information, they can work alongside mission control to monitor space systems and alert astronauts if something requires their attention. Some people even dream of allowing robots to perform less complicated spacewalks; as NASA's Robonaut2 has demonstrated, in isolated situations a robot may be able to throw switches or do some of the more routine tasks of porting stuff around.

To be fair, robots largely do the heavy lifting on the International Space Station anyway; between Canadarm2 and Dextre, much of the maintenance is done under the careful control of technicians at the CSA. In the far future, however, robots will likely be more autonomous and will provide more real-time information about spacecraft systems and humans. It seems not too far away to imagine astronauts popping pills with ingestible sensors embedded with rich capabilities to monitor the human body, or working alongside computers that provide coaching about what to do next on a deep-space mission.

Early 2022 is replete with discussions about the "metaverse" and the implications for our daily life on Earth. Beyond individual new immersive technologies, such as virtual reality and AI, the metaverse is a complex network of 3D virtual worlds, a hypothetical integration of social media, physical spaces, avatars and even digital economies. (In a November 2021 article, *WIRED* magazine said, "To a certain extent, talking about what 'the metaverse' means is a bit like having a discussion about what 'the internet' means in the 1970s. The building blocks of a new form of communication were in the process of being built, but no one could really know what the reality would look like."[33]) Already incorporated to some degree in massively multiplayer online video games enabled by the internet, metaverse technologies are the latest example of science fiction becoming science fact, a phenomenon not unusual in the brave new world of space exploration.

It wasn't that long ago that the "Father of Electronics," Lee de Forest, stridently declared, "Man is inherently an earthly creature, and only his scientific imagination will ever make him a planetary emigrant." De Forest invented the vacuum tube that made possible radio and television broadcasting as well as radar technology; despite holding over 300 patents worldwide, he had a tumultuous career and many commercial failures. He also got one prediction terribly wrong. In February 1957, the Associated Press reported that de Forest told an audience listening to Voice of America that "to place a man in a multi-stage rocket and project him into the controlling gravitational field of the moon, where the passenger can make scientific observations, perhaps land alive, and then return to Earth — all that constitutes a wild dream worthy of Jules Verne. . . . I am bold enough to say that such a man-made moon voyage will never occur regardless of all future scientific advances."[34] De Forest died on June 30, 1961, too early to witness the first Moon landing just over eight years later on July 20, 1969.

In thinking about what is possible for future space exploration, it is far more inspiring to remember the words of President John F. Kennedy when he challenged the May 25, 1961, joint session of Congress to dream big, words that still ring true today: "Now it is time to take longer strides . . . which in many ways may hold the key to our future here on Earth. . . . I believe that this nation should commit itself to achieving the goal, before this decade is out, of landing a man on the Moon and returning him safely to Earth. No single space project in this period will be more impressive to mankind, or more important for the long-range exploration of space. . . . This gives the promise of some day providing a means for even more exciting and ambitious exploration of space, perhaps beyond the Moon, perhaps to the very end of the solar system itself."[35]

CHAPTER 16

Exploring Beyond

T his book has described the space medicine journey that humans have taken in the first 60 or so years of space exploration. The majority of these voyages have been in low Earth orbit, and a select few — a handful of crews during the Apollo program — made it to the Moon. As of this writing, NASA and numerous international partners are planning a second attempt at voyaging to the Moon, this time establishing long-term bases there, with an eye to eventually (perhaps in the 2030s or the 2040s) moving on to Mars. The old joke about exploring the Red Planet is that it will always happen 20 years from now — because that has consistently been the target, announced in the 1960s, the 1980s, the 2000s and even now in the 2020s.

Predicting the future of space travel is complex, which is true of just about any technology project. Even the most visionary people do occasionally stumble when making predictions. Microsoft founder Bill Gates once said he couldn't imagine why Gmail had such a large storage limit, imagining that people would always depend on storage space on a computer rather than in a cloud,[1] and there are numerous other folks who thought computers would never shrink beyond the size of a "single room," to quote the movie *Apollo 13*. Even further back in time, we can find predictions from Western Union in 1876 that the telephone had no future,[2] from Lord Kelvin in 1878 saying that X-rays are "clearly a hoax,"[3] and perhaps the most infamous bad prediction of all time: the *New York Times* proudly proclaiming that flight would be possible "from one million to ten million years" after 1903 — this written just two months before the Wright brothers showed heavier-than-air flight was indeed possible.[4] The venerable newspaper also showed that it is not afraid to make apologies, however, as it did in July 1969 — just as the Apollo 11 crew was about to fly to the Moon — atoning for a January 1920 article that said, in part, that rockets couldn't possibly fly in space.[5] Oops.

Perhaps when we talk about the future of space travel, we talk about it knowing that we might be just plain wrong about it. Funding for space programs has historically been highly volatile and dependent on such things as a country remaining together (lessons of the Soviet Union), an urgent and perceived national need propelling a hugely expensive program (lessons of NASA's Apollo years), and a presidential administration committing to continuing work on a predecessor's program for landing on the Moon or Mars (which has hardly ever happened, although Richard Nixon sort of did it and Joe Biden appears to be doing it).

We also can't forget the acceleration of space technology in the last few decades. People once laughed at Elon Musk for dreaming of private spaceships and self-landing rockets, but in 2021 we watched his spacecraft soar into space — although the safety issues will ultimately determine how effective a venture SpaceX will be at bringing astronauts aloft in the long term. Private spaceflight is indeed accelerating, with Axiom Space and SpaceX planning trips to the International Space Station and beyond the Moon, so it may be that a private coalition of some sort goes to Mars rather than a government program. That is, unless China happens to play out some of the Biden administration's worst fears and works to one-up the United States on Moon landings and Mars landings — much as the U.S. feared of the Soviet Union in the 1950s and 1960s.

Despite the difficulty of learning from the past and making predictions, we can start to talk about what *may* be possible for humanity in the next 100 years or even in the next 1,000 years, given current medical knowledge and our current knowledge of physics. Any of this might be overturned quickly, and there remain many questions about how far humans will have travelled in space and where humans might be in our solar system at the end of this millennium. One thing that always fires the imagination when thinking about the future is turning to science fiction, and there are many favorites — *Star Trek*, *Star Wars*, *The Expanse*, *Battlestar Galactica*, *Interstellar*. Authors such as Arthur C. Clarke and Isaac Asimov often come up as visionaries to think about in the 2020s, and we hope the future of human spaceflight will be even more inclusive of different cultures and ideas. For now, the facts in science and health will guide missions beyond Earth orbit to create a better understanding of the challenges humans face in exploring the far universe.

The next 100 years are a little easier to imagine. It isn't too much of a stretch to talk about folks moving around the solar system, perhaps — *perhaps* — regularly visiting such easy-to-reach destinations as the Moon or Mars. "Easy" is of course a relative term, but the Moon and Mars are rocky bodies that are within a short journey from Earth; at this time, the Moon is only about three days away, and Mars six to nine months away when the orbits of Earth and Mars align. However, this is based on using the current chemically propulsive technology. Nuclear propulsion, as NASA points out, may cut down travel time within the solar system considerably, allowing a round-trip Mars mission to complete in as little as two years, instead of the current three. Naturally, a shorter travel time would reduce many of the health problems associated with long-term exploration, including radiation dosage and the bone/muscle/organ changes that occur with long periods of microgravity. NASA has a current radiation limit of 50 mSv/year, or 50 rem/year,[6] with specific lifelong radiation limits for individual astronauts determined by age and gender (women's lower body mass may make them more vulnerable). But that is for low Earth orbit and would need to be re-evaluated based on the realities of a Martian trip, where larger amounts of radiation would be present on the surface due to the planet's lack of a magnetic field. Then again, radiation shielding might improve considerably in the coming decades due to technology developments.

Another asset of nuclear propulsion, the experts say, is that using it would seriously cut down on propellant use compared with chemical rockets,[7] leaving less to haul to Mars and possibly fewer rockets or smaller rockets for the long trip. "Nuclear electric propulsion systems use propellants much more efficiently than chemical rockets but provide a low amount of thrust," NASA wrote in 2021.[8] "They use a reactor to generate electricity that positively charges gas propellants like xenon or krypton, pushing the ions out through a thruster, which drives the space-craft forward. Using low thrust efficiently, nuclear electric propulsion systems accelerate spacecraft for extended periods and can propel a Mars mission for a fraction of the propellant of high thrust systems."

Indeed, a 2021 report from NASEM, the National Academies of Sciences, Engineering, and Medicine, suggested some form of nuclear propulsion would be best for repeatable and affordable human trips to Mars.[9] There are other options, Ars Technica reported: nuclear thermal propulsion, where the nuclear reactor would burn liquid hydrogen as a fuel; or nuclear electric propulsion, which transforms heat from a fission reactor to electrical power that then accelerates an ionized propellant like xenon.[10]

Naturally, the report reminds us, this technology won't appear with a snap of the fingers and NASA will need to start investing quickly to make nuclear a feasible option in the 2030s. "So far, the agency has been somewhat reticent to move quickly on nuclear propulsion," Ars Technica added. "This may be partly because the space agency is so heavily invested in the Space Launch System rocket and chemical propulsion needed for the Artemis moon program."[11]

That said, NASA's possible reluctance may be an interpretation rather than a reality. President Joe Biden had only taken office weeks before February 2021, when nuclear propulsion drew the attention of the media with the release of the new NASEM report. Also, for years NASA has been facing ongoing restrictions in its supply of plutonium-238 because of the United States' commitment in the 1980s to nuclear disarmament protocols. The Perseverance rover was only able to go to Mars on nuclear power after Oak Ridge National Laboratory made the first plutonium-238 in a generation in 2015, which NASA paid for.[12]

Nuclear propulsion appears to be a leading, vital technology for short-range exploration of the solar system, and in situ resource utilization (ISRU) is another that is frequently cited. As of April 2020, NASA was pursuing ISRU — more simply put, "living off the land" — to allow the Artemis program to go forward at what the agency hopes will be a reduced cost, in the long term. As the agency stated, "Rovers will carry a variety of instruments including ISRU experiments that will generate information on the availability and extraction of usable resources (e.g., oxygen and water). Advancing these technologies could enable the production of fuel, water, and/or oxygen from local materials, enabling sustainable surface operations with decreasing supply needs from Earth."[13]

Simply put, NASA intends to scout out appropriate resources on the Moon and then to use them to support human missions. Already we have a sense of the resources available thanks to the Lunar Reconnaissance Orbiter, which has been mapping the Moon since 2009, detecting deposits of minerals and ice, among other features. The next challenge will be getting some "ground truth," as the geology experts term it, to see how much of each resource is indeed available and accessible; it is presumed that most lunar ice is present near the surface in deep craters sheltered from the sun. Rovers and landers are a cheaper way to survey the Moon first to see what might be there, and then human missions can be developed with an eye to mining those resources. NASA has a Commercial Lunar Payload Services program in place to encourage private landers and rovers to explore the surface, and as of this writing the first of these landings may take place in the early 2020s, with Astrobotic's Peregrine Mission 1. Further flights are imagined later in the 2020s that would bring a suite of different instruments to the Moon to assess its potential for water and other valuable resources.

Similarly on Mars, space explorers may be able to make use of what appear to be underground reserves of water; indeed, a March 2021 journal article in *Science* suggested that between 30 and 99 percent of the water supply of Mars from its early history might be buried underground,[14] with the rest having vanished into space as the planet's atmosphere thinned over millennia. The thinning was likely due to solar radiation striking the atmosphere, which is unprotected by a magnetic field, and stripping lighter molecules away — a phenomenon that the ongoing Mars Atmosphere and Volatile Evolution mission (MAVEN) is exploring.[15] "Water can break down rocks through a process called chemical weathering, which some-times results in minerals becoming hydrated," the Verge technology blog reported at the time.[16] "Hydrated minerals take up and store water, locking it away. For example, gypsum, a water-soluble mineral found naturally on Mars, can keep its water trapped unless heated at temperatures higher than 212 degrees Fahrenheit."

The big question is whether future Martian explorers will have to do deep drilling underground, as data from both NASA's Mars Reconnaissance Orbiter and Europe's Mars Express don't give much indication about how much rock is in the way of any underground reserves. "You have to wave your hands and extrapolate about how thick that layer is that you see at the surface," Michael Meyer, the lead scientist for NASA's Mars exploration program, told the Verge.

Worse, plumbing the depths of Mars is tougher than we imagined. NASA struggled for the better part of two Earth years to get the InSight mission's drill a few feet below the surface, implementing innovative methods such as gently hammering on it with a robotic arm to try to get past some unexpectedly stubborn soil. NASA eventually called it quits in January 2021, explaining: "The soil's unexpected tendency to clump deprived the spike-like mole of the friction it needs to hammer itself to a sufficient depth. After getting the top of the mole about two or three centimeters under the surface, the team tried one last time to use a scoop on InSight's robotic arm to scrape soil onto the probe and tamp it down to provide added friction. After the probe conducted 500 additional hammer strokes on Saturday, Jan. 9, with no progress, the team called an end to their efforts."[17]

The Moon and Mars appear to be the most feasible destinations for long-term exploration in the next 100 years, but there are a few other potential destinations. For a while, the Obama administration tasked NASA with sending astronauts to an asteroid, possibly in the 2020s, to pursue an asteroid redirect test to assist with planetary defense.[18] While the mission died shortly after the arrival of the Trump administration in 2016, humans may be able to visit asteroids that are relatively close to the orbit of Earth, perhaps even in the vicinity of Mars or the asteroid belt at farthest.

Working in the extremely low gravity of an asteroid will be a considerable challenge; it may be more like a spacewalk outside the space station, as the astronauts would work their way around the asteroid rather than walk directly on it. Such challenges would also be true of the tiny, asteroid-like moons of Mars, Phobos and Deimos, if humans were to expand the surface exploration of the Red Planet to include them. More speculatively, perhaps one day astronauts will travel to visit the clouds of Venus[19] — a relatively calm and friendly environment compared with its hell-like surface. Or they may go in the opposite direction, travelling farther out in the solar system to the surface of Europa,[20] one of the many moons of Jupiter and a focus of NASA's Europa Clipper mission — although how to bring humans to Europa, which has intense radiation, is unknown at this time. Studies from

NASA and others in the scientific community are still at an early stage right now, making it difficult to effectively evaluate the possibilities of sending humans to these destinations in the next century or so.

To tread further into the future and farther into the universe, there is a fine line between science and the realm of science fiction, most famously embodied in the *Star Trek* and *Star Wars* franchises. Fundamentally, the problem is trying to traverse long distances. The nearest known planetary system is roughly four light-years away from Earth, which would be roughly 1,000 years of travel using the relatively speedy nuclear propulsion.[21] It therefore appears that venturing beyond the solar system will require a fundamental rethinking of exploration. We're either going to have to shorten the journey in some way, or we're going to require humans to survive a lengthy voyage. There are numerous possibilities to explore in both directions, many worthy of another book entirely. Let's briefly look at two of the most popular ideas in each category — warp drive and wormholes for shortening the time, and hibernation or multigenerational voyages to deal with the possibility that we'll be in space for a while.

WARP DRIVE OR HYPERDRIVE

Star Wars and *Star Trek* are both famous for using faster-than-light travel, appearing to make their starships or spaceships move swiftly at the touch of the button or the pull of a lever. Faster-than-light is also a staple of other franchises like *Battlestar Galactica*, whose 2004–2009 series showed the difficulties of pinpoint navigation across space — first, you're in a hurry (for much of the series they are being chased by humanoid robots known as Cylons) and second, a lot of your ships are not true battleships but are carrying settlers. But in the true sense, would a warp drive work?

A 2021 article in The Conversation[22] reminds us of research by Mexican theoretical physicist Miguel Alcubierre in 1994, who showed that compressing "spacetime" — essentially, the mathematical model familiar to Einstein followers that encompasses the three dimensions of space and the fourth dimension of time — is theoretically possible under Einstein's General Relativity. Unfortunately, there's a problem in the way, which is the warp drive would depend on generating a considerable amount of negative energy.

We don't mean that feeling of malaise that comes when you encounter a bad day at work or at home. Rather, it has to do with how objects attract each other in space using gravity; it is difficult to get too far into the details as they quickly stray into a discussion of quantum mechanics rather than medicine, but a quick excerpt from The Conversation illustrates the issue.

"To create negative energy, a warp drive would use a huge amount of mass to create an imbalance between particles and antiparticles," the article states.[23] The authors go on to discuss the role of negatively charged particles called electrons and their opposite in quantum theory, known as antielectrons: "If an electron and an antielectron appear near the warp drive," the article continues, "one of the particles would get trapped by the mass and this results in an imbalance. This imbalance results in negative energy density. Alcubierre's warp drive would use this negative energy to create the spacetime bubble. But for a warp drive to generate enough negative energy, you would need a lot of matter. Alcubierre estimated that a warp drive with a 100-meter bubble would require the mass of the entire visible universe."

Now, these are mathematical models and scientists are not at the point of putting a dog in a starship and rocketing it to high speeds to see if it will work (to borrow a bit from the time travel of Back to the Future). If future advances in engineering can actually make warp drive, hyperdrive or faster-than-light travel available, that would open up the solar system to us. Perhaps there's another solution, though.

WORMHOLES

Another staple of science fiction is the wormhole, which is a hypothetical connection between vastly separated areas of space. You don't have to look far to find references to this theory. The children's story A Wrinkle in Time includes an excellent discussion of how wormholes might work, essentially imagining an ant on a piece of string whose ends are brought together for the ant not to have to walk as far. Wormholes have also been used in movies such as Contact (1997), based on the 1984 book by Carl Sagan, which portrayed wormhole travel to the star Vega, some 26 light-years from Earth. Fans of Stargate, Babylon 5, Doctor Who, Crusade and the Marvel movies will also recognize them, and even Star Trek had a wormhole next to the station "Deep Space 9." The science of wormholes was again portrayed in a recent space movie hit, *Interstellar* (2014), and there are undoubtedly more examples. Although they are highly fictional, there is some research on these phenomena by well-known physicists such as Kip Thorne on the preprint website arXiv.[24]

A 2014 Space.com article starts with the limitations in the theory of wormholes: "Though wormholes are a favored sci-fi trope, nobody knows whether or not they actually exist. According to Einstein's theory of general relativity, they are possible, but no sign of them has ever been spotted. Furthermore, scientists say, a wormhole would likely collapse quickly unless it was propped open using some kind of negative-energy matter. So the big wormhole in 'Interstellar' would require some serious and exotic engineering work," wrote Mike Wall.[25]

That said, Thorne provided some mathematical equations to the filmmakers to help make the idea of wormholes believable, even though they may not necessarily work in real life. "I think the laws of physics very probably forbid warp drives and traversable wormholes," Thorne told *Scientific American* in 2014.[26] "The research that has gone on over the past 25 years trying to determine whether it's possible all point in negative directions, but it's not a firmly closed door."

He defended the use of wormholes in *Interstellar* on grounds familiar to space fans, which is that fiction is not necessarily *supposed* to be true to reality. "To a great extent, my motivation here was to try to use the movie as a lure to get people who might otherwise not have much interest in science curious about it, by exposing them to strange, exotic phenomena like wormholes," Thorne said. "The film is the bait, and the book [Thorne's *The Science of Interstellar*] is the hook I want to use to draw them in even further, to get them to dig in and learn something new. If they are young, maybe they will consider careers in science rather than in finance or law. If they are older, I still think it's tremendously important that larger fractions of our citizenry possess enough understanding of science to appreciate its powers and its limitations."

Ultimately, wormholes and warp drive may not be practical solutions to transport people swiftly through vast distances in the universe, mostly because it seems math stands in the way of making it biologically possible. But what about if we were to take the space medicine route and to think about two other common solutions — hibernation and multigenerational travel? We've discussed hibernation earlier in the book and while it seems promising, the time it would take to travel to destinations farther out in our solar system present a challenge for today's nascent technology. It took the Voyager 2 spacecraft travelling at 42,000 mph twelve years to get to Neptune[27] and contemplating a return trip would mean roughly 25 years traveling in a spacecraft — notwithstanding the time spent visiting a gas planet that's made of a thick fog of water, ammonia and methane surrounding an Earth-sized solid center. Perhaps the great distances to reach such destinations requires a multigenerational approach.

MULTIGENERATIONAL VOYAGES

Multigenerational voyages are a perennial theme of science fiction dating back roughly 100 years.[28] Simply put, the notion of multigenerational voyages assumes that you build a spacecraft sturdy enough to survive decades or perhaps hundreds of years of travel, and then you fill it with people who are prepared to live their lives, and perhaps even die, aboard the ship. The group would include couples producing children, who in turn would carry on the voyage and perhaps go on to have their *own* children, and so forth, until the ship makes it all the way to its destination or accomplishes a return trip.

These ideas raise some fundamental questions about mental health and human will. As much as the first generation of people may be willing to go and be healthy enough to undertake the endeavor, they would essentially be forcing their children to continue the journey. What would happen if the next generation of voyagers — or even the third or fourth or fifth generation — wanted to turn around and go back to Earth, assuming of course that some postapocalyptic scenario isn't at play on the Blue Planet? Earth may be too far away and supplies insufficient for the ship to do a turn-around voyage.

There are also some obvious engineering questions regarding sustainability. It would be enough of a challenge to establish and support a space settlement on another world where we know beforehand there would be access to resources that may generate water, oxygen, structural materials (through regolith) and the like. But sending a starship on a one-way voyage to a completely unfamiliar world would require an extraordinary degree of robustness; there are simply far too many unknowns to make it practically feasible. Losing oxygen would be simply catastrophic as there may be no way to regenerate it. The same goes for lack of water — a tough resource to find in the more distant regions of space — or food, which would simply not be available through resupply or foraging en route.

Such a voyage may not be impossible, but there are fundamental logistical difficulties that must be overcome, let alone the engineering and technology challenges. The current International Space Station missions show some of the issues that may arise over six months or a year of travel and how space agencies and astronauts can overcome them. But the long-term effects of radiation, low gravity or microgravity, and the other hazards of space travel are no trivial matter.

Clearly there are numerous obstacles to overcome before humankind ventures farther out into the universe, but there are some other ways that we may be able to explore the outer reaches of space without subjecting humans to the inherent dangers. Already robots are allowing us to experience faraway places virtually, gradually sending information back to Earth as bandwidth allows. Or perhaps there will be some future advancement in telescopes we cannot yet foresee, allowing us to view unimagined detail on planets and other worldscapes far away from us.

It is no small wonder that humankind is so enthralled with the unknowables of space; but for all the technological advances that can aid in space exploration, the human body will be one of the most interesting and enriching assets to work with as we explore the universe. While limited in so many ways, the body is the window through which we perceive and experience the world, and other worlds, around us. The 1966 film *Fantastic Voyage*[29] attempted to take viewers on "the most incredible adventure that man could ever achieve," as a miniaturized submarine with its crew aboard is injected into a man's body to remove a deadly blood clot in his brain. The trailer described this inner journey as "One of the miracles of the universe. . . . Exploring an unknown universe, unknown dangers," sounding remarkably like the wonder we experience in space exploration. Perhaps the most daring, inspiring voyage we can take is into the inner world of the human body — infinitely mysterious and fascinating in its workings, adaptations and creativity. And only by understanding and exploring our inner landscape can we gain the knowledge, experience and inspiration to explore other worlds and the mysteries of the cosmos.

ACKNOWLEDGMENTS

Once again, I would like to thank my co-author Elizabeth Howell for her hard work, enthusiasm and support. This is the second book we have co-authored during the pandemic — when we wrote *Leadership Moments from NASA: Achieving the Impossible*, it was just starting. And now that we are finishing this book it is hopefully ending. To me, exploration is the relentless quest for knowledge. This philosophy has guided the course of my career as a neuroscientist, physician, astronaut and aquanaut, and arguably it is the underlying theme of this book. The lessons learned from the COVID-19 pandemic can provide insights into the importance of planetary protection that will hopefully be remembered to help control the risk and impact of future pandemics. This is the ninth pandemic that has taken place during my career as a physician and unfortunately, they will recur in the future. The best approach we can take is to learn from the past, embrace scientific knowledge and research, and develop new approaches to their prevention. The impact of infectious disease is an ever-present reminder that we live in a global village — what happens in one part of the world can affect all of us. It is an opportunity to collaborate in search of solutions.

As an undergraduate student at McGill University in the seventies, I spent many hours studying in the Osler Library of the History of Medicine. Surrounded by books written by some of the greatest clinicians in history, I dreamt of what the future might bring — the possibility of a career as a physician. One of my textbooks was the *Textbook of Medical Physiology* by Arthur C. Guyton and John E. Hall. It is one of the classic texts in medical education, and my 1971, blue-covered edition was well used. As an undergraduate, I had yet to learn the importance of focus in preparing for tests and I often found myself drawn to section VIII of the book that dealt with aviation, space and deep-sea diving physiology rather than studying the course material. One of the memorable moments in my career was sharing that story with the authors after becoming an astronaut and aquanaut. Now my tenth edition of the text is prominently displayed on my office bookshelf, signed by Drs. Guyton and Hall.

I hope our readers enjoy this book and the interesting aspects of comparative physiology that are part of understanding how humans adapt to living in space. We have tried to share a blend of unique stories with the historic experience and clinical lessons learned by the first generation of experts in space medicine. It will be exciting to see the future of space medicine grow to support a new era of human space exploration.

To the NASA Chief Medical Officers, Drs. Arnauld Nicogossian, Rich Williams and J.D. Polk, it was a pleasure working with you. Thanks for being excellent mentors, colleagues and great friends. I would also like to thank my colleagues in the former NASA Space and Life Sciences Directorate at Johnson Space Center — Drs. John Rummel, Larry Dietlein, Craig Fischer, Sam Pool and Jeff Davis — for your commitment to furthering the field of space medicine. Special thanks to all the former and current NASA, international partner and commercial flight surgeons for your relentless commitment to caring for astronauts before, during and after their missions.

It has been a privilege to work with experts in space medicine globally in furthering human space exploration, and it has been a special honor to have worked with my Canadian colleagues Drs. Gary Gray, Joan Saary, Jean-Marc Comtois, Bob Thirsk, Roberta Bondar, David Saint-Jacques, Doug Hamilton and Andy Kirkpatrick to ensure Canada is at the forefront of this exciting field. To the team at Defence Research and Development Canada, home of Canadian aerospace and diving medicine, thanks for everything you do to keep the Canadian Forces and astronauts safe while pushing the edge of the envelope.

I would also like to thank the team at Leap Biosystems for their relentless passion in developing new technologies to deliver healthcare everywhere.

The past year has been incredibly busy as we worked together integrating the payload for the flight of Canadian private astronaut Mark Pathy on the Axiom Aerospace's historic first commercial mission to the ISS. It was a unique opportunity to support this mission and a very proud moment for the Leap team to have their MedChecker© technology onboard. Leap was also thrilled to have worked with AEXA Aerospace to demonstrate two-way holoportation between the ISS and mission control at Johnson Space Center. It was exciting when Mark Pathy and I became the first humans to experience two-way holoportation between the Earth and space! This unique technology promises to change the way we provide behavioral support to astronauts in space and in collaboration with the Leap Biosystem's AR technologies, transform the future of space medicine. Special thanks to the private astronauts, the NASA teams at Marshall Spaceflight Center and Johnson Space Center, the Canadian Space Agency, as well as the Axiom and AEXA teams for making this mission a reality.

Once again, I truly appreciate the support and enthusiasm of our publisher Jack David and the ECW team. Thanks for your excitement about this project, the unwavering encouragement and the many hours reviewing the manuscript. It is greatly appreciated.

To my wife, Cathy, and Olivia, Evan and Theo, thanks for your support and understanding with my long hours at the keyboard.

DAVID R. WILLIAMS OC OONT BSC MSC MD

CM FCFP FRCPC FRCP LLD (HON) DSC (HON)

My co-author, Dave Williams, has been a guiding light through this manuscript through his vision for delivering space medicine to the benefit of others. He has a unique gift for making the complex easy to understand by bringing in everyday examples to illustrate the themes of medicine that everyone should know. I am grateful that our second book was such a fruitful relationship; I have gained so much from his experience and his work.

Dave alluded to living through nine pandemics during his career. I can only speak to the experience of seeing one up close, which was when non-COVID-19 pneumonia brought me to emergency wards at the Montfort

Hospital and the Ottawa Hospital, General Campus in September 2021. Doctors, nurses, ambulance attendants and other front-line staff gave the most professional care, doing everything to make me feel comfortable and safe during a moment when health care facilities were overwhelmed by a large wave of COVID-19 cases. Honestly, I may not have been here today without their care. As a society, we are very lucky to have these front-line workers, and I only hope our society will give them the support they need.

COVID-19 taught me other lessons as well, namely about precarity and partnerships. As a freelancer, I saw businesses I work for taking all sorts of approaches to deal with financial precarity. I want to thank every last one of you who doubled down to keep interesting work flowing to me when so much else in the world was a strange blur. On the partnership side, as a resident of Ottawa, I saw friends and colleagues unite for the benefit of their neighbors and local charities during a nasty three-week occupation of Ottawa in February 2022 that eventually triggered federal intervention. I did my best to pitch in, but really, they showed me what it means to be truly involved in making a community equitable, diverse and inclusive. The pandemic has thus shaped much of how I think these latter years, and it also was a large driver in writing this book as I saw writing as a form of community service.

Dave and I both realize that books don't exist without readers, and I thank all of you for taking the time to read our words and for supporting an industry that has been battered by supply chain issues due to COVID-19. Please do continue to support your local bookstores when you can afford to do so — and to get books like this on the shelf of your local library for your communities.

ECW was the publisher that launched my book-writing career (*Canadarm and Collaboration*, published in 2020 but accepted back in 2018 when I was a much less "known" individual). There are so many editors, marketers, designers and helpers involved in the process that it is impossible to thank every one of you. That said, I do want to especially thank Karen Milner (of Milner Associates, a contractor of ECW) for her patience during an odd moment in my life that unfortunately collided with a big deadline, and to ECW's Samantha Chin for helping to marshal a manuscript through the strange schedules of the co-authors. I also owe Jack David huge thanks for accepting a third book proposal and bringing me through two more years of working with an incredible team.

As a freelancer and a person with too many degrees, I've had a lot of positive people and influences through my career and only have space to (unfairly) pick a few. Everyone at Carleton University's journalism program (bachelor's) and the University of North Dakota's space studies department (masters and PhD) — thanks for educating me. To Athabasca University, where I expect to finish a bachelor's degree in history in 2022 — thanks for being around for people like me with competing obligations and for launching a probably lifelong interest in United Kingdom archaeology. And to the team that made the Hollywood movie *Apollo 13*, released in 1995 — thanks for launching my interest in space when I was a young teenager when I caught the end of it on VHS at my middle school in June 1996.

Thanks also to every single person who has ever hired me for a story, for editing, for teaching or for anything that brought me income or experience. I want to especially single out Tariq Malik at Space.com and Marc Boucher at SpaceQ for getting me launched into the business and for still asking for my writing after more than 10 years together. I'm also grateful to the Professional Development Institute at the University of Ottawa for helping me deepen my teaching practice for so many people and to the various post-secondary institutions that have hired me for contracts over the years. You all keep me going in so many ways, and my job now is to give back to others as much as possible.

Special thanks to my virtual assistant, Christina Goodvin, for providing feedback on early drafts and assisting with the editing process. Your assistance was particularly crucial in keeping our bibliography updated, which was no easy task — I am very grateful.

Lastly, thanks to all the friends and family who have patiently heard me talk about the inevitable sticky points of book-writing — and were kind enough not to remind me I've done this often before and should know it comes with the territory. My husband, J, is a supporter in so many ways, both morally and as a willing household manager when things get busy. Also, thanks to our cat (Gabriel) and our dog (Geordi) for willingly hanging out with me after a long day typing things.

ELIZABETH HOWELL BJOUR MSC PHD

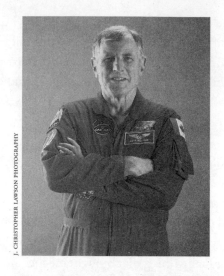

DR. DAVE WILLIAMS is an astronaut, aquanaut, pilot, emergency physician, scientist and CEO. He has flown to space twice, once on the space shuttle Columbia and once on Endeavour, logging over 13 million kilometers in space and over 17 hours of spacewalks. He holds the Canadian spacewalking record for the most spacewalks and is the first Canadian to have lived and worked on the world's only undersea research habitat. He is the recipient of six honorary degrees, the Order of Canada and the Order of Ontario.

He was selected by the Canadian Space Agency in 1992 and became the manager of the missions and space medicine group the following year. After his first spaceflight, he was chosen to be the Director of the Space & Life Sciences Directorate at the Johnson Space Center in Houston, Texas, while concurrently serving as the Deputy Associate Administrator of the Office of Spaceflight at NASA Headquarters. Since retiring from the space program, he was the CEO at Southlake Regional Health Centre and is currently the CEO of Leap Biosystems.

He has written six books, including his memoir, *Defying Limits: Lessons from the Edge of the Universe*, and co-authored a highly acclaimed book entitled *Leadership Moments from NASA: Achieving the Impossible*. Through his experiences as a physician, astronaut, senior executive and CEO, he is a highly respected speaker and consultant in leadership, teamwork, risk management, organizational agility and resilience. He lives in Toronto, ON.

DANIELLE DONDERS

ELIZABETH HOWELL, PHD, is a staff writer for Space.com. She has been a space journalist for 20 years and is one of the few Canadian journalists to focus exclusively on space.

After working for the *Globe and Mail*, the *Canadian Medical Association Journal*, CTV Ottawa and the *Ottawa Business Journal*, Howell struck out as a full-time freelance space journalist in 2012. Before shifting to full-time work at Space.com in 2022, she was a frequent contributor to publications such as Space.com, *Forbes*, CBC and the Canadian astronomy magazine *SkyNews*. Her on-site reporting includes five human spaceflight launches (from Florida and Kazakhstan) and embedded journalism from a simulated Red Planet mission at the Mars Society's Mars Desert Research Station near Hanksville, Utah.

Howell has also been a communications instructor for nearly 10 years at more than half a dozen educational institutions, including Algonquin College, University of Ottawa and La Cité. She creates writing and instructional materials focused on communicating complex information to multiple audiences while respecting equity, diversity and inclusion. She received degrees at Carleton University (Bachelor of Journalism) and the University of North Dakota (Masters of Space Studies, PhD in Aerospace Sciences.) She expects to complete her Bachelor of History at Athabasca University in 2022, following a lifelong passion for learning about the past.

Why Am I Taller? is Howell's fifth book. She lives with her husband in Ottawa.

NOTES

CHAPTER 1

1. NASA, "Christina Koch Returns to Earth," YouTube video, 1:16, February 6, 2020, https://www.youtube.com/watch?v=K9ctQAdw-DA.
2. Ibid.
3. NASA, "Hear from Record-Breaking NASA Astronaut Christina Koch," YouTube video, 31:57, February 12, 2020, https://www.youtube .com/watch?v=UncCrD1cGb4.
4. Press Association, "Tim Peake Nursing 'World's Worst Hangover' after Six Months in Space," *Guardian*, June 20, 2016, https://www.theguardian .com/science/2016/jun/20/tim-peake-nursing-worlds-worst-hangover-after -six-months-in-space.
5. "'Gravity Is Not My Friend,' Canadian Astronaut David Saint-Jacques Says as He and Crewmates Return," CBC News, June 24, 2019, https://www.cbc.ca/news/technology/david-saint-jacques-1.5187999.
6. Allison Koehler, "Back to Earth — A Medical Perspective," *Alexander Gerst's Horizons Blog*, European Space Agency (ESA), December 14, 2018, https://blogs.esa.int/alexander-gerst/2018/12/14/back-to-earth-a -medical-perspective/.
7. Gary Jordan, "The Zero-G Workout," February 20, 2018, interviews Andrea Hanson in *Houston We Have a Podcast*, episode 33, transcript, last

updated February 23, 2018, https://www.nasa.gov/johnson/HWHAP /the-zero-g-workout.

8. "Mission Discovery: Launching C.O.L.B.E.R.T. into Space," Space. com, March 2, 2010. https://www.space.com/7872-mission-discovery -launching-space.html.

9. "Do Tread on Me," NASA ISS Behind-the-Scenes, August 19, 2009, https://www.nasa.gov/mission_pages/station/behindscenes/colbert _feature.html.

10. "Exercising in Space," *Exploring Space Through MATH*, student edition, 2, https://www.nasa.gov/pdf/516063main_Alg_ST_Exercising-in-Space %2012-23-10.pdf.

11. Jordan, "Zero-G Workout," (see chap. 1, n. 7).

12. R. Donald Hagan, et al., "Musculoskeletal Effects of 16-Weeks of Training with the Advanced Resistive Exercise Device (aRED) (ROI _ARED)," Life Sciences Data Archive, n.d., https://lsda.jsc.nasa.gov/ Experiment/exper/1304.

13. "Kinesiology Researchers Work with Astronauts to Study Aging and Cardiovascular Health," University of Waterloo, June 21, 2019, https:// uwaterloo.ca/kinesiology-health-sciences/news/kinesiology-researchers -work-astronauts-study-aging-and.

14. Tariq Malik, "NASA Astronaut Completes Boston Marathon in Space," Space.com, April 16, 2007, https://www.space.com/3702-nasa-astronaut -completes-boston-marathon-space.html.

15. Robert Z. Pearlman, "British Astronaut Tim Peake Sets Off-World Record Running Marathon in Space," Space.com, April 25, 2016, https:// www.space.com/32682-tim-peake-breaks-space-marathon-record.html.

16. Jordan, "Zero-G Workout," (see chap. 1, n. 7).

17. Ibid.

18. Denise Chow, "Giffords' Astronaut Brother-in-Law Speaks Out on Tragedy," Space.com, January 18, 2011, https://www.space.com/10643 -scott-kelly-discusses-tucson-shooting-space-station.html.

19. Dan Huot, "Resident Extreme," December 20, 2018, interviews Christina Koch in *Houston We Have a Podcast*, episode 82, transcript, last updated March 8, 2019, https://www.nasa.gov/johnson/HWHAP /resident-extreme.

20. NASA, "Christina Koch" (see chap. 1, n. 3).

21. Bin Wu, et al., "On-Orbit Sleep Problems of Astronauts and Countermeasures," *Military Medical Research* 5, no. 1 (May 30, 2018): 17, https://www.ncbi.nlm.nih.gov/pmc/articles/PMC5975626/.

22. Melissa Gaskill, "Let There Be (Better) Light," NASA, October 19, 2016, https://www.nasa.gov/mission_pages/station/research/let-there-be -better-light.

23. Monica Edwards and Laurie Abadie, "NASA's Twins Study Results Published in Science Journal, April 11, 2019, NASA, https://www.nasa .gov/feature/nasa-s-twins-study-results-published-in-science.

24. Ibid.

CHAPTER 2

1. Melissa Gaskill, "International Space Station Research Keeps an Eye on Vision Changes in Space," NASA, August 3, 2020, https://www .nasa.gov/mission_pages/station/research/news/iss-20-evolution-of -vision-research.

2. Kelli Mars, "What Is Spaceflight Associated Neuro-Ocular Syndrome?" NASA, July 29, 2021, https://www.nasa.gov/image-feature/what-is -spaceflight-associated-neuro-ocular-syndrome.

3. Elizabeth Usovicz, "Are You Defying Gravity? Newton's Law for Salespeople," Thinking Bigger, November 4, 2014, https://ithinkbigger .com/defying-gravity-newtons-law-salespeople/.

4. David P. Barash, "How Necking Shaped the Giraffe," *Nautilus*, May 21, 2015, https://nautil.us/issue/24/error/how-necking-shaped-the -giraffe?curator=Informerly.

5. Sarah Treleaven, "Defying Gravity," *University of Toronto Magazine*, December 8, 2010, https://magazine.utoronto.ca/campus/history /photo-wilbur-franks-g-suit-pilot-uniform/.

6. Barash, "Giraffe" (see chap. 2, n. 4).

7. "Supported Headstand," *Yoga Journal*, August 28, 2007, https://www .yogajournal.com/poses/supported-headstand/.

8. "22 Facts Regarding the Mind | World Brain Day," MickySays, last updated August 18, 2021, https://mickysays.com/22-facts-regarding -the-mind/.

9. Rinad S. Minvaleev, et al., "Headstand (Sirshasana) Does Not Increase the Blood Flow to the Brain," *Journal of Alternative and Complementary Medicine* 25, no. 8 (June 2019): 827–832.

10. Guy S. Reeder, et al., "Use of Doppler Techniques," Mayo Clinic Proceedings, September 1, 1986, https://www.mayoclinicproceedings .org/article/S0025-6196(12)62774-8/fulltext.

11. Minvaleev, "Headstand (Sirshasana)" (see chap. 2, n. 9).

12. Mani Baskaran, et al., "Intraocular Pressure Changes and Ocular Biometry during Sirsasana in Yoga Practitioners," *Ophthalmology* 113 (August 2006): 1327–1332.

13. Giovanni Taibbi, et al., "Ocular Outcomes Evaluation in a 14 Day Head-Down Bed Rest Study," *Aviation, Space, and Environmental Medicine* 85, no. 10 (October 2014): 983–92.

14. Alex S. Huang, et al., "Gravitational Influence on Intraocular Pressure," *Journal of Glaucoma* 28, no. 8 (August 2019): 756–764.

15. Scott H. Greenwald, et al., "Intraocular Pressure and Choroidal Thickness Respond Differently to Lower Body Negative Pressure during Spaceflight" *Journal of Applied American Physiological Society* 131, no. 2 (August 2021): 613–620.

CHAPTER 3

1. "Air-to-Ground and Onboard Voice Tape Transcription," NASA, Mission Transcript: Gemini VIII, https://historycollection.jsc.nasa .gov/JSCHistoryPortal/history/mission_trans/gemini8.htm.

2. Retro Space HD, "Gemini VIII — Described by the Crew — Neil Armstrong — First Crewed Docking — March 1966," YouTube video, 11:39, February 16, 2020, https://www.youtube.com/watch?v =999bm7PcZ_0.

3. Ibid.

4. S.R. Mohler, "Tumbling and Spaceflight: the Gemini VIII Experience," *Aviation, Space, and Environmental Medicine* 61, no. 1 (January 1990): 62–6.

5. Ibid.

6. "Spinning, Rolling, and Swinging! Oh My!" Penn State Extension,

Better Kid Care, 2017, https://extension.psu.edu/programs/betterkidcare/news/2017/spinning.

7. Stephen T. Moore, et al., "Long-Duration Spaceflight Adversely Affects Post Landing Operator Proficiency," *Scientific Reports* (February 2019), https://www.nature.com/articles/s41598-019-39058-9.

8. Sarah Jane Bell, "What Happens to an Astronaut's Ability to Drive after a Space Mission?" ABC, February 25, 2019, https://www.abc.net.au/news/2019-02-26/would-you-trust-an-astronaut-to-drive-your-car/10708998.

9. William H. Paloski, et al., "Risk of Sensory Motor Performance Failures Affecting Vehicle Control during Space Missions: A Review of the Evidence," *Journal of Gravitational Physiology* 15, no. 2 (December 2008), https://humanresearchroadmap.nasa.gov/Evidence/reports/sensorimotor.pdf.

10. ESA, "Tim Peake's Dizziness Experiment," YouTube video, 3:36, June 12, 2016, https://www.youtube.com/watch?v=2Lz5UeROyXM.

11. Charles A. Berry and J.L. Homick, "Findings on American Astronauts Bearing on the Issue of Artificial Gravity for Future Manned Space Vehicles," *Aerospace Medicine* 44, no. 2 (1973): 163–168.

12. "What Causes Sea Sickness," National Oceanic and Atmospheric Administration, last updated October 6, 2021, https://oceanservice.noaa.gov/facts/seasickness.html.

13. Richard Boyle, et al., "Neural Re-adaptation to Earth's Gravity Following Return from Space," in *The Neurolab Spacelab Mission: Neuroscience Research in Space* (Houston, TX: NASA, 2003), 45–49.

14. Gary Larson, *Beyond the The Far Side* (FarWorks, Inc., 1983).

CHAPTER 4

1. "Valiant (episode)," Memory Alpha Fandom, Wiki, n.d., https://memory-alpha.fandom.com/wiki/Valiant_(episode).

2. "Earl Grey Tea," Memory Alpha Fandom, Wiki, n.d., https://memory-alpha.fandom.com/wiki/Earl_Grey_tea.

3. "What Really Is Astronaut Food?" Smithsonian National Air and Space Museum, November 8, 2021, https://airandspace.si.edu/stories/editorial/what-really-astronaut-food.

4. John Uri, "Space Station 20th: Food on ISS," August 14, 2020, https:// www.nasa.gov/feature/space-station-20th-food-on-iss.

5. Smithsonian, "Astronaut Food" (see chap. 4, n. 3).

6. Uri, "Food on ISS" (see chap. 4, n. 4).

7. Smithsonian, "Astronaut Food" (see chap. 4, n. 3).

8. NASA, "Food for Space Flight," February 26, 2004, https://www.nasa. gov/audience/forstudents/postsecondary/features/F_Food_for_Space _Flight.html.

9. Uri, "Food on ISS" (see chap. 4, n. 4).

10. Ibid.

11. "Train Like an Astronaut: Taste in Space," NASA, Educator Section, HREC CB 2102014, n.d., 1–6, https://www.nasa.gov/sites/default/files /files/Taste-in-space-TLA-FINAL.pdf.

12. Alan Boyle, "Poop in Space Revisited: Apollo 10's Floating Turds Pop Up 44 Years Later," NBC News, April 9, 2013, https://www.nbcnews.com /sciencemain/poop-space-revisited-apollo-10s-floating-turds-pop-44-years -1c9284102.

13. Ibid.

14. Peng Jiang, et al., "Reproducible Changes in the Gut Microbiome Suggest a Shift in Microbial and Host Metabolism during Spaceflight," *Microbiome* 7, no. 113 (August 2019), https://microbiomejournal. biomedcentral.com/articles/10.1186/s40168-019-0724-4.

15. Alexander A. Voorhies, et al., "Study of the Impact of Long-Duration Space Missions at the International Space Station on the Astronaut Microbiome," *Scientific Reports* (July 2019), https://www.nature.com /articles/s41598-019-46303-8.

16. Elizabeth Rajan, "Digestion: How Long Does It Take?" Mayo Clinic, December 2019, https://www.mayoclinic.org/digestive-system/expert -answers/faq-20058340

17. Lakshmi Putcha, "Gastrointestinal Function During Extended Duration Space Flight (DSO 622)," Life Sciences Data Archive, n.d., https://lsda.jsc.nasa.gov/Experiment/exper/621.

18. Scott M. Smith, et al., *Human Adaptation to Spaceflight: The Role of Nutrition*, 9, https://www.nasa.gov/sites/default/files/human -adaptation-to-spaceflight-the-role-of-nutrition.pdf.

19. Lakshmi Putcha, "Bed Rest Standard Measures: Actilight (BRSMAct),"

Life Sciences Data Archive, n.d., https://lsda.jsc.nasa.gov/Experiment /exper/1280.

20. Anand Narayan, "Simulated Microgravity-Induced Systemic Inflammation and Its Impact on Circulatory Function and Structure (Postdoctoral Fellowship) (80NSSC19K0426)," Life Sciences Data Archive, n.d., https://lsda.jsc.nasa.gov/Experiment/exper/18005.

21. Smith, 30 (see chap. 4, n. 18).

22. Ibid, 30.

23. Ibid, 60.

24. Ibid, 61.

25. "Water Recycling," NASA, October 13, 2014, https://www.nasa.gov /content/water-recycling/.

26. Eric M. Jones, "Debrief and Goodnight," Apollo 16 Lunar Surface Journal, 1997, last updated May 12, 2014, https://www.hq.nasa.gov /alsj/a16/a16.debrief1.html.

27. Harold M. Schmeck Jr., "Apollo Crew to Get More Restricted Potassium Diet," *New York Times* December 5, 1972, https://www.nytimes.com /1972/12/05/archives/apollo-crew-to-get-more-restricted-potassium-diet .html.

28. William Paloski, "How Humans Adapt to Spaceflight: Physiological Changes," in NASA's *Wings In Orbit*, 2019, 371–407, https://www .nasa.gov/centers/johnson/pdf/584739main_Wings-ch5d-pgs370-407 .pdf.

29. NASA, CSA & Impact Canada Initiative, *Deep Space Food Challenge: Phase 1 Competition Rules*, 2021, https://static1.squarespace.com/static /5fd5ab003c1f6275809f31d9/t/60c7cf10b738762702d8f839/1623707485771 /FNL_NASA_DSF_Phase_1_Rules-rev-1.pdf.

30. Ibid.

31. G.W. Wieger Wamelink, et al., "Can Plants Grow on Mars and the Moon: A Growth Experiment on Mars and Moon Simulants," *PLOS One*, August 27, 2014, https://journals.plos.org/plosone/article?id=10.1371 /journal.pone.0103138.

32. Kevin M. Cannon and Daniel T. Britt, "Feeding One Million People on Mars," *New Space* 7, no. 4 (December 2019), https://www .liebertpub.com/doi/abs/10.1089/space.2019.0018?journalCode=space.

33. "MOXIE," NASA Science, Mars 2020 Mission Perseverance Rover, spacecraft instruments summary, n.d., https://mars.nasa.gov/mars2020 /spacecraft/instruments/moxie/.

34. Cannon, "Feeding One Million" (see chap. 4, n. 32).

35. Charles Q. Choi, "How to Feed a Mars Colony of 1 Million People," Space.com, September 18, 2019, https://www.space.com/how-feed -one-million-mars-colonists.html.

CHAPTER 5

1. Melanie Whiting, ed., "Skylab: America's First Space Station," NASA, May 14, 2018, https://www.nasa.gov/feature/skylab-america-s-first -space-station.

2. Mark Garcia, ed., "Skylab 2: Mission Accomplished!", NASA, June 22, 2018, https://www.nasa.gov/feature/skylab-2-mission-accomplished.

3. Gerald P. Carr, interviewed by Kevin M. Rusnak, JSC Oral History Project, October 25, 2000, https://historycollection.jsc.nasa.gov /JSCHistoryPortal/history/oral_histories/CarrGP/CarrGP_10-25-00.htm.

4. John Uri, "The Real Story of the Skylab 4 'Strike' in Space," NASA, November 16, 2020, https://www.nasa.gov/feature/the-real-story-of -the-skylab-4-strike-in-space.

5. Henry S.F. Cooper, "Life in a Space Station-II," *The New Yorker*, September 6, 1976, https://www.newyorker.com/magazine/1976/09/06 /life-in-a-space-station-ii.

6. Paul Vitello, "William Pogue, Astronaut Who Staged a Strike in Space, Dies at 84," *New York Times*, March 10, 2014, https://www.nytimes .com/2014/03/11/science/space/william-r-pogue-astronaut-who-flew -longest-skylab-mission-is-dead-at-84.html.

7. Tony Schwartz and Catherine McCarthy, "Manage Your Energy, Not Your Time," *Harvard Business Review*, October 2007, https://hbr.org /2007/10/manage-your-energy-not-your-time.

8. David J. Eicher, "Rusty Schweickart Remembers Apollo 9," *Astronomy*, March 2019, https://astronomy.com/magazine/news/2019/03/rusty -schweickart-remembers-apollo-9.

9. Nathan A. Cranford and Jennifer L. Turner, "Spaceflight Standard Measures: Characterizing How Humans Adapt in Space," NASA, January 11, 2021, https://www.nasa.gov/feature/spaceflight-standard -measures-characterizing-how-humans-adapt-in-space.

10. Ibid.

11. Melissa Gaskill, "Let There Be (Better) Light," NASA, October 19, 2016, https://www.nasa.gov/mission_pages/station/research/let-there -be-better-light.

12. Scott Kelly, "When you live on the International Space Station, you work weekends," LinkedIn, October 27, 2017, https://www.linkedin .com/pulse/when-youre-astronaut-international-space-station-you-work -scott-kelly/.

13. "An Astronaut's Work," NASA, May 27, 2004, https://www.nasa.gov /audience/forstudents/9-12/features/F_Astronauts_Work.html.

14. Ibid.

15. Radiogram No. 7162u, NASA, November 12, 2014, https://www.nasa .gov/sites/default/files/files/111214_tl.pdf.

16. "Ed Lu's Journal: Entry #9: Day in the Life," SpaceRef.com, August 17, 2003, http://www.spaceref.com/news/viewsr.html?pid=10038.

17. Ibid.

18. Kelly, "International Space Station" (see chap. 5, n. 12).

19. Suzanne Bell and Stephen T. Vander Ark, "Behavioral Health," NASA, February 23, 2015, https://www.nasa.gov/content/behavioral-health.

20. "HI-SEAS: Hawaii Space Exploration Analog and Simulation," https://www.hi-seas.org/.

21. C. Heinicke, et al., "Crew Self-Organization and Group-Living Habits during Three Autonomous, Long-Duration Mars Analog Missions," *Acta Astronautica* 182 (May 2021): 160–178, https://www.sciencedirect .com/science/article/pii/S0094576521000606.

22. Elizabeth Howell, "Nasa's Perseverance Rover Team Will Have to Live on 'Mars Time' after Landing on the Red Planet," Space.com, February 18, 2021, https://www.space.com/perseverance-rover-mission-on-mars -time.

23. SpaceRef.com, "Ed Lu's Journal" (see chap. 5, n. 16).

CHAPTER 6

1. Jeff Mangum, "Why Is the Earth's Orbit Around the Sun Elliptical?" National Radio Astronomy Observatory (NRAO), https://public.nrao .edu/ask/why-is-the-earths-orbit-around-the-sun-elliptical/.

2. "Mars in Our Night Sky," NASA Science, Mars Exploration Program, https://mars.nasa.gov/all-about-mars/night-sky/close-approach/.

3. "Cruise," NASA Science, Mars 2020 Mission Perseverance Rover, https://mars.nasa.gov/mars2020/timeline/cruise/.

4. David R. Williams, "A Crewed Mission to Mars," original page author — M. J. Shaw, NASA Goddard Space Flight Center, last updated November 16, 2015, https://nssdc.gsfc.nasa.gov/planetary/mars/marsprof .html.

5. Emma Barrett, *Extremes: Why Some People Thrive at the Limits*, first edition, (New York: Oxford University Press, 2014), 67.

6. Eldora Valentine, "Race From Space: Suni Williams Runs Boston Marathon," Phys.org, April 16, 2007, https://phys.org/news/2007-04 -space-suni-williams-boston-marathon.html.

7. "Do Bears Actually Hibernate?" Science World, February 19, 2016, https://www.scienceworld.ca/stories/do-bears-actually-hibernate/.

8. Ibid.

9. Arielle Emmett, "Sleeping Their Way to Mars," *Air & Space Magazine*, April 2017, https://www.airspacemag.com/space/hibernation-for-space -voyages-180962394/.

10. Mike Wall, "Hibernating Astronauts May Be Key to Mars Colonization," Space.com, August 30, 2016. https://www.space.com/33894-mars -colonization-hibernating-astronauts-torpor.html.

11. John Bradford, Spaceworks Engineering Inc., "Advancing Torpor Inducing Transfer Habitats for Human Stasis to Mars," May 13, 2016, https://www.nasa.gov/feature/advancing-torpor-inducing-transfer -habitats-for-human-stasis-to-mars/.

12. Ibid.

13. Christine Aschwanden, "Wanted: Mars Explorers. Must Be Able To Tolerate Boredom and Play Nice With Others," FiveThirtyEight, September 30, 2016, https://fivethirtyeight.com/features/wanted-mars -explorers-must-be-able-to-tolerate-boredom-and-play-nice-with-others/.

14. Megan Gannon, "How NASA Will Keep Astronauts From Going Stir-Crazy on Long Space Missions," NBC News, March 13, 2017, https://www.nbcnews.com/mach/space/how-nasa-preparing-astronauts-minds-long-mars-mission-n732711.

15. Denise Chow, "6 Mock Mars Explorers Emerge from 520-Day Virtual Mission," Space.com, November 4, 2011. https://www.space.com/13500-mock-mars-mission-mars-500-ends.html.

16. ESA, "Mars 500: One Year Inside," YouTube video, 4:35, May 30, 2011, https://www.youtube.com/watch?v=nM_fmLxzqhQ.

17. Associated Press, "Mars Mission Simulation Ends after 520 Days," CBC, November 4, 2011, https://www.cbc.ca/news/science/mars-mission-simulation-ends-after-520-days-1.1096114.

18. Geoffrey York, "Canadian's Harassment Complaint Scorned," *The Globe and Mail*, March 25, 2000. https://www.theglobeandmail.com/news/national/canadians-harassment-complaint-scorned/article25458615/.

19. Mathias Basner, et al., "Mars 520-D Mission Simulation Reveals Protracted Crew Hypokinesis and Alterations of Sleep Duration and Timing," *Proceedings of the National Academy of Sciences* (PNAS), February 12, 2013, https://www.pnas.org/content/110/7/2635.

20. Nadia Drake, "Here's What It Feels Like to Spend a Year on 'Mars,'" *National Geographic*, August 29, 2016, https://www.nationalgeographic.com/science/article/nasa-mars-hi-seas-hawaii-human-mission-space-science.

21. Katerina Stepanova, "Virtual Reality Can Combat Isolation with Awe and Empathy — on Earth and in Space," The Conversation, October 31, 2021, https://theconversation.com/virtual-reality-can-combat-isolation-with-awe-and-empathy-on-earth-and-in-space-170189.

22. Marina Koren, "When a Mars Simulation Goes Wrong," *The Atlantic*, June 22, 2018, https://www.theatlantic.com/science/archive/2018/06/mars-simulation-hi-seas-nasa-hawaii/553532/.

CHAPTER 7

1. "American Cockroach," Wikipedia, n.d., https://en.wikipedia.org/wiki/American_cockroach.

2. "Cockroach Facts," Western Exterminator, n.d., https://www
.westernexterminator.com/cockroaches/cockroach-facts/.

3. Graham N. Askew, et al., "Limitations Imposed by Wearing Armour
on Medieval Soldier's Locomotor Performance," Royal Society, July
20, 2011, https://doi.org/10.1098/rspb.2011.0816.

4. Jason R. Norcross, et al., *Feasibility of Performing a Suited 10-km
Ambulation on the Moon - Final Report of the EVA Walkback Test
(EWT)*, NASA, TP-2009-214796, October 2009, https://lsda.jsc.nasa
.gov/lsda_data/dataset_inv_data/EWT__34015388.pdf_EPSP_EWT
_2012_53_040216.pdf.

5. Mae Mills Link, "Space Medicine in Project Mercury," NASA,
SP-4003, 1965, p. 7.

6. Alexei Leonov, "The Nightmare of Voskhod 2," *Air & Space Magazine*,
January 2005, https://www.smithsonianmag.com/air-space-magazine
/the-nightmare-of-voskhod-2-8655378/.

7. Ibid.

8. Ibid.

9. David R. Williams, NASA, EVA Debrief, STS-118, 2007.

10. Michael Neufeld, "Almost Blind and Completely Exhausted: Gene
Cernan's Disastrous Gemini Spacewalk," Smithsonian National Air
and Space Museum, June 4, 2011, https://airandspace.si.edu/stories
/editorial/almost-blind-and-completely-exhausted-gene-cernans
-disastrous-gemini-spacewalk.

11. Bob Granath, "Gemini XII Crew Masters the Challenges of
Spacewalks," NASA Kennedy Space Center, November 14, 2016,
https://www.nasa.gov/feature/gemini-xii-crew-masters-the-challenges
-of-spacewalks.

12. Ibid.

13. Richard Hollingham, "Apollo in 50 Numbers: Time," BBC Future,
July 19, 2019, https://www.bbc.com/future/article/20190718-apollo
-in-50-numbers-time.

14. Hamilton Sundstrand, "Space Shuttle Extravehicular Mobility Unit
(EMU) Life Support Subsystem (LSS) and Space Suit Assembly (SSA)
Data Book," Revision II, September 2002.

15. "The Astronaut Diaper," ISS National Laboratory, April 1, 2016,
https://www.issnationallab.org/stem/lesson-plans/astronaut-diaper/.

16. "Even Homes in Space Need a Door," NASA Science, July 5, 2001, https://science.nasa.gov/science-news/science-at-nasa/2001/ast06jul_1.

17. "Space Station Spacewalks," NASA, mission pages, January 19, 2022. https://www.nasa.gov/mission_pages/station/spacewalks/.

18. Tariq Malik, "Damaged Spacesuit Glove Ends ISS Spacewalk Early," Space.com, August 15, 2007, https://www.space.com/4226-damaged-spacesuit-glove-ends-iss-spacewalk-early.html.

19. E. Christiansen, et al., "MMOD Risk Assessments for EVA," *Orbital Debris Quarterly News* 19, no. 3 (July 2015), https://orbitaldebris.jsc.nasa.gov/quarterly-news/pdfs/odqnv19i3.pdf,

20. Miriam Kramer, "Spacesuit Leak That Nearly Drowned Astronaut Could Have Been Avoided," Space.com, February 26, 2014, https://www.space.com/24835-spacesuit-water-leak-nasa-investigation.html.

21. Dava Newman, "Building the Future Spacesuit," *Ask Magazine*, 37–40, https://www.nasa.gov/pdf/617047main_45s_building_future_spacesuit.pdf.

22. Dava Newman, "How to Create a Spacesuit," TED Archive, August 29, 2017, YouTube video, 11:08, https://www.youtube.com/watch?v=lZvP_URAjmM.

23. Lex Fridman, "Dava Newman: Space Exploration, Space Suits, and Life on Mars," *Lex Fridman Podcast*, Apple Podcast #51, 2019, https://lexfridman.com/dava-newman/.

CHAPTER 8

1. Jan Riley, "The Keratin Trilogy: Skin, Hair and Nails," Ausmed, https://www.ausmed.com/cpd/articles/what-is-keratin.

2. Jeanna Bryner, "Why Do We Have Fingernails?" Live Science, February 7, 2013, https://www.livescience.com/32472-why-do-we-have-fingernails.html.

3. Ibid.

4. Deb M. Eldredge, "Hoof Anatomy: What Horse Hooves Are Made Of," Horseman's Report, HorseHealthProducts.com, May 8, 2018, https://www.horsehealthproducts.com/horsemans-report/hoof-leg-care/hoof-anatomy.

5. Mariah Adcox, "How Fast Do Nails Grow? Contributing Factors and Tips for Growth," Healthline, April 13, 2018, https://www.healthline.com/health/beauty-skin-care/how-fast-do-nails-grow.

6. W.B. Bean, "A Discourse on Nail Growth and Unusual Fingernails," *Transactions of the American Clinical and Climatological Association* 74 (1962): 152–67, PMID: 14044604, https://pubmed.ncbi.nlm.nih.gov/14044604/.

7. Seriously Science, "This Professor Measured His Fingernail Growth for 35 Years. The Results Will Amaze You!" *Discover*, April 13, 2015, https://www.discovermagazine.com/health/this-professor-measured-his-fingernail-growth-for-35-years-the-results-will-amaze-you.

8. "Nail Splitting (Onychoschizia)," Skinsight, November 26, 2018, n.d., https://www.skinsight.com/skin-conditions/adult/onychoschizia.

9. Ibid.

10. George McGavin, "The Incredible Human Hand and Foot," BBC News, February 18, 2014, https://www.bbc.com/news/science-environment-26224631.

11. James Taylor, "Lesson 1b: 'Nails 101' (HD) — Official James Taylor Guitar Lessons," YouTube video, 5:44, December 16, 2010, https://www.youtube.com/watch?v=7BqISqpMR08.

12. The Classical Guitar Store, "Fingernails and the Classical Guitar," n.d., http://www.classicalguitarstore.com/fingernails/.

13. Melissa Malamut "The Best Nail Cream Is Meant for Horses," *The Strategist*, October 21, 2016, https://nymag.com/strategist/2016/10/best-nail-strengthener-for-healthy-strong-nails.html.

14. *National Geographic*, "Skin," January 18, 2017, https://www.nationalgeographic.com/science/article/skin-1.

15. Benjamin Radford, "Is House Dust Mostly Dead Skin?" Live Science, December 11, 2012, https://www.livescience.com/32337-is-house-dust-mostly-dead-skin.html.

16. Megan Garber, "The Disgusting Side of Space: What Happens to Dead Skin in Microgravity," *The Atlantic*, August 1, 2013, https://www.theatlantic.com/technology/archive/2013/08/the-disgusting-side-of-space-what-happens-to-dead-skin-in-microgravity/278274/.

17. Caleb Wong, "How to Shower in Space," Smithsonian National Air and

Space Museum, July 18, 2017, https://airandspace.si.edu/stories/editorial/how-shower-space.

18. Ibid.

19. insideISS, "ISS Science Garage — Eeewwww," YouTube video, 2:38, July 30, 2013, https://www.youtube.com/watch?v=akw6XBjD5ho&t=14s.

20. Garber, "The Disgusting Side of Space" (see chap. 8, n. 16).

21. David Caballero, "The Types of Microorganisms That Live on Our Skin," Skin Trust Club, April 9, 2021, https://www.skintrustclub.com/blog/the-types-of-microorganisms-that-live-on-our-skin/.

22. Luke K. Ursell, et al., "Defining the Human Microbiome," *Nutrition Reviews* 70, suppl. 1 (August 2012): S38–S44, https://www.ncbi.nlm.nih.gov/pmc/articles/PMC3426293/.

23. NASA, "Space Station Leaves 'Microbial Fingerprint on Astronauts," Phys.org, May 14, 2020, https://phys.org/news/2020-05-space-station-microbial-fingerprint-astronauts.html.

24. Ibid.

25. Ibid.

26. Carrie Gilder, "Study Finds Space Station Microbes Are No More Harmful Than Those Found in Similar Ground Environments," NASA, International Space Station Program Science Office, Johnson Space Center, January 22, 2020, https://www.nasa.gov/mission_pages/station/research/news/space-station-microbes-no-more-harmful-than-those-on-earth-extremophiles.

27. Ibid.

28. Sylvain Louradour, Netexplo Observatory, *The New Now*, no. 1 (2021): 172, https://netexplo.com/static/media/uploads/pdf/the_new_now_eng_pdf_book.pdf.

CHAPTER 9

1. Jeanna Bryner, "First Known Case of Coronavirus Traced Back to November in China," Live Science, March 14, 2020, https://www.livescience.com/first-case-coronavirus-found.html.

2. "Ingredients for Life?" NASA Europa Clipper, https://europa.nasa.gov/europa/life-ingredients/.

3. "NASA's Hubble Spots Possible Water Plumes Erupting on Jupiter's Moon Europa," NASA, September 26, 2016, https://www.nasa.gov /press-release/nasa-s-hubble-spots-possible-water-plumes-erupting-on -jupiters-moon-europa.

4. "Enceladus: Ocean Moon," NASA Science, last updated September 25, 2018, https://solarsystem.nasa.gov/missions/cassini/science/enceladus/.

5. Marc Kaufman, "Large Reservoir of Liquid Water Found Deep Below the Surface of Mars," Astrobiology at NASA, July 25, 2018, https:// astrobiology.nasa.gov/news/large-reservoir-of-liquid-water-found-deep -below-the-surface-of-mars/.

6. Tom Metcalfe, "Hidden Beneath a Half Mile of Ice, Antarctic Lake Teems with Life," Live Science, January 15, 2019, https://www.livescience .com/64501-buried-lake-antarctica-life.html.

7. "China's Zhurong Rover Lands on Mars (Update: New Images)," Sky & Telescope, May 17, 2021, https://skyandtelescope.org/astronomy-news /chinas-zhurong-rover-lands-on-mars/.

8. "Mars 2020 Perseverance Launch Press Kit," NASA Jet Propulsion Laboratory, https://www.jpl.nasa.gov/news/press_kits/mars_2020/launch /mission/spacecraft/biological_cleanliness/.

9. "Planetary Protection," NASA Office of Safety & Mission Assurance, last modified December 5, 2021, https://sma.nasa.gov/sma-disciplines /planetary-protection.

10. Ibid.

11. John D. Rummel, et al. (eds.), *A Draft Test Protocol for Detecting Possible Biohazards in Martian Samples Returned to Earth*, NASA Technical Reports Server, (Moffett Field, CA: Ames Research Center, October 2002): 1, https://ntrs.nasa.gov/api/citations/20030053046 /downloads/20030053046.pdf.

12. Ibid, 1.

13. Ibid, 5.

14. Ibid, 7.

15. Leonard David, "Could Mars Samples Brought to Earth Pose a Threat to Our Planet? What the Coronavirus (and 'Andromeda Strain') Can Teach Us," Space.com, April 8, 2020, https://www.space.com/mars-sample -return-threat-earth-coronavirus-andromeda-strain.html.

16. Donald L. Savage, et al., "Meteorite Yields Evidence of Primitive Life

on Early Mars," NASA, August 7, 1996, https://www2.jpl.nasa.gov/snc
/nasa1.html.

17. "President Clinton Statement Regarding Mars Meteorite Discovery,"
    Office of the Press Secretary, August 7, 1996, https://www2.jpl.nasa.gov
    /snc/clinton.html.

18. Savage, "Meteorite Yields Evidence" (see chap. 9, n. 16).

19. "Carbon Compounds from Mars Found Inside Meteorites," NASA,
    August 8, 1996, https://www.nasa.gov/mission_pages/mars/multimedia
    /pia00289.html.

20. Craig Covault, "Three Martian Meteorites Triple Evidence for Mars
    Life," Spaceflight Now, January 9, 2010, https://spaceflightnow.com/news
    /n1001/09marslife/.

21. Richard S. Johnston and John A. Mason, "Chapter 1: The Lunar
    Quarantine Program," in *SP-368: Biomedical Results of Apollo*, 1975,
    https://history.nasa.gov/SP-368/s5ch1.htm.

22. Elizabeth Howell, "Apollo 11 Moon Landing Had a Plan for Lunar
    Germs — But Video Clip Reveals a Big Flaw," Space.com, July 6,
    2019, https://www.space.com/pbs-chasing-the-moon-apollo-11
    -quarantine.html.

23. Johannes Kemppanen, "Apollo Lunar Quarantine: A 50th Anniversary
    View," *Apollo Flight Journal*, NASA, last updated July 22, 2019, https://
    history.nasa.gov/afj/lrl/apollo-quarantine.html.

24. Ibid.

25. Ibid.

26. Ibid.

27. Johnston and Mason, "Lunar Quarantine Program" (see chap. 9, n. 21).

28. Ibid.

29. Ibid.

30. John Uri, "50 Years Ago: Apollo 13 and German Measles," NASA,
    April 2, 2020, https://www.nasa.gov/feature/50-years-ago-apollo-13
    -and-german-measles.

31. "Apollo 13," NASA, July 8, 2009, https://www.nasa.gov/mission_pages
    /apollo/missions/apollo13.html.

32. NASA, *Flight Crew Health Stabilization Program: Space Medicine
    Division, Revision D*, November 2010, https://ntrs.nasa.gov/api/citations
    /20130000048/downloads/20130000048.pdf.

33. Ibid, Section 7.
34. Ibid, Section 7.2.
35. Ibid, Section 7.3.1.
36. Ibid, Section 7.3.4.
37. Ibid, Section 7.3.6.
38. Victor Glover (@AstroVicGlover), Twitter post, May 8, 2021, 2:00 p.m., https://twitter.com/AstroVicGlover/status/1391090708396875782.
39. National Academies of Sciences, Engineering, and Medicine, *Assessment of the Report of NASA's Planetary Protection Independent Review Board* (Washington, DC: The National Academies Press, 2020), https://www.nap.edu/catalog/25773/assessment-of-the-report-of-nasas-planetary-protection-independent-review-board.
40. Ker Than, "Stanford's Scott Hubbard Contributed to New 'Planetary Quarantine' Report Reviewing Risks of Alien Contamination," *Stanford News*, May 7, 2020, https://news.stanford.edu/2020/05/07/new-planetary-quarantine-report-reviews-risks-alien-contamination-earth/.
41. Ibid.
42. Ibid.

CHAPTER 10

1. "In the Region of Lost Minds," *New York Times*, archived review, https://archive.nytimes.com/www.nytimes.com/books/98/12/06/specials/sacks-mistook.html.
2. Oliver Sacks, *The Man Who Mistook His Wife for a Hat* (New York: Harper Perennial, HarperCollins Publishers, 1985), 45.
3. Fabrice Sarienga, et al., "Proprioception, Our Imperceptible Sixth Sense," Medical Press, Neuroscience, The Conversation, February 17, 2021, https://medicalxpress.com/news/2021-02-proprioception-imperceptible-sixth.html.
4. Ibid.
5. Sacks, 47 (see chap. 10, n. 2).
6. Paul Chrustowski, "Living Without Proprioception," interview with Ian Waterman, Worth Publishers, YouTube video, 4:05, October 28, 2015, https://www.youtube.com/watch?v=pMEROPOK6v8.

7. James Lackner, "Some Proprioceptive Influences on the Perceptual Representation of Body Shape and Orientation," Brain III (Pt. 2) (2): 281–97. May 1, 1998.

8. Mel Slater, "The Pinocchio Effect," Faculty of Psychology, University of Barcelona, YouTube video, 2:58, July 7, 2021, https://www.youtube.com/watch?v=7GgQeBznC4I.

9. Otmar Bock, et al., "Visual-Motor Coordination During Spaceflight," in *The Neurolab Spacelab Mission: Neuroscience Research in Space* (Houston, TX: NASA, 2003), 83.

10. Ibid, 85.

11. Ibid, 86.

12. Carol Stock Kranowitz, "3 Types of Sensory Disorders That Look Like ADHD," ADDitude, April 21, 2021, https://www.additudemag.com/slideshows/signs-of-sensory-processing-disorder/.

13. Ibid.

14. Laura Di Orio, "Why Are Dancers So Clumsy?" Dance Informa, https://www.danceinforma.com/2013/11/05/why-are-dancers-so-clumsy/.

15. Phoebe Woei-Ni Hwang, Kathryn L. Braun, "The Effectiveness of Dance Interventions to Improve Older Adults' Health: A Systematic Literature Review," *Alternative Therapies in Health Medicine* 21, no. 5 (2015): 64–70.

16. Clara Moskowitz, "Buzz Aldrin Gears Up For 'Dancing With the Stars,'" Space.com, March 9, 2010, https://www.space.com/8019-buzz-aldrin-gears-dancing-stars.html.

17. Liz Mineo, "The Balance in Healthy Aging," *Harvard Gazette*, April 25, 2017, https://news.harvard.edu/gazette/story/2017/04/tai-chi-can-prevent-elderly-from-falls-add-mental-agility/.

CHAPTER II

1. John Glenn and Nick Taylor, *John Glenn: A Memoir*, Bantam (NY: 2000), 473.

2. Ibid, 473–4.

3. Ibid, 474.

4. John Noble Wilford, "John Glenn, American Hero of the Space Age, Dies at 95," *New York Times*, December 8, 2016, https://www.nytimes.com/2016/12/08/us/john-glenn-dies.html.

5. Lawrence K. Altman, "Studying Aging in Space? Send an Aging Astronaut," *New York Times*, January 27, 1998, https://www.nytimes.com/1998/01/27/science/studying-aging-in-space-send-an-aging-astronaut.html.

6. Abigail Trafford, "John Glenn: Hero, Guinea Pig," *Washington Post*, October 13, 1998, Z06, https://www.washingtonpost.com/wp-srv/national/longterm/glenn/stories/heromed.htm.

7. Altman, "Aging in Space" (see chap. 11, n. 5).

8. John H. Glenn Jr., interviewed by Sheree Scarborough, JSC Oral History Project, August 25, 1997, https://historycollection.jsc.nasa.gov/JSCHistoryPortal/history/oral_histories/GlennJH/GlennJH_8-25-97.htm.

9. Valerie Neal, "John Glenn's Return to Space on Discovery," Smithsonian National Air and Space Museum, October 29, 2018, https://airandspace.si.edu/stories/editorial/john-glenns-return-space-discovery.

10. "Active NASA astronauts in their 60s", collectSPACE.com, October 10, 2020, http://www.collectspace.com/ubb/Forum38/HTML/002348.html.

11. Altman, "Aging in Space" (see chap. 11, n. 5).

12. D.J. Dijk, et al., "Sleep, Performance, Circadian Rhythms, and Light-Dark Cycles during Two Space Shuttle Flights," *American Journal of Physiology-Regulatory, Integrative and Comparative Physiology* 281, no. 5 (2001): R1647-R1664, https://pubmed.ncbi.nlm.nih.gov/11641138/.

13. "CCISS: How Changes in Gravity Affect the Heart," Canadian Space Agency (CSA), last modified December 12, 2017, https://www.asc-csa.gc.ca/eng/sciences/cciss.asp.

14. "The Vascular Series: Studying Heart Health in Space," CSA, last modified January 7, 2020, https://www.asc-csa.gc.ca/eng/sciences/vascular.asp.

15. Sharon Aschaiek, "A New Campus Residence Opens Just for Seniors," January 20, 2016, *University Affairs*, https://www.universityaffairs.ca/news/news-article/new-campus-residence-opens-just-for-seniors/.

16. CSA, "Vascular Series" (see chap. 11, n. 14).

17. David Coulombe, "Audio — Interview With Dr. Richard Hughson," transcript, last modified November 23, 2015, https://web.archive.org /web/20211019110416/https://cihr-irsc.gc.ca/e/49523.html.

18. CSA, "Vascular Series" (see chap. 11, n. 14).

19. Coulombe, "Richard Hughson" (see chap. 11, n. 17).

20. Leroy Chiao, "Return to Earth: An Astronaut's View of Coming Home," March 1, 2016, https://www.space.com/32102-astronaut-perspective-on -long-deployments.html.

21. "Ten Ways Astronaut Christina Koch Will Need to Readjust to Earth After 328 Days in Space," NASA, February 3, 2020, https://www.nasa .gov/feature/ten-ways-astronaut-christina-koch-will-need-to-readjust -to-earth-after-328-days-in-space/.

22. Ibid.

23. Denise Chow, "Astronaut Jessica Meir's Return to Earth Has Been Far from Ordinary," NBC News, April 30, 2020, https://www.nbcnews .com/science/space/astronaut-jessica-meir-s-return-earth-has-been-far -ordinary-n1196671.

24. "Dressing Astronauts for Return to Earth," NASA, March 28, 2019, https://www.nasa.gov/mission_pages/station/research/news/b4h-3rd/hh -dressing-astronauts-for-return.

25. "Return and Conditioning," CSA, last modified September 26, 2019, https://www.asc-csa.gc.ca/eng/resources-young/junior-astronauts /activities/fitness-and-nutrition/return-and-reconditioning.asp.

26. Bruce Nieschwitz, et al., "Post Flight Reconditioning for US Astronauts Returning from the International Space Station," NASA Technical Reports Server, 2011, https://ntrs.nasa.gov/api/citations /20110020318/downloads/20110020318.pdf.

27. Ron Berler, "Inside the Crazy Hardcore Fitness Regimen That Keeps American Astronauts Healthy and Strong in Space," *Men's Journal*, https://www.mensjournal.com/health-fitness/inside-crazy-hardcore -fitness-regimen-keeps-american-astronauts-healthy-and/.

28. Ibid.

29. Ibid.

30. "The Human Body in Space," NASA, February 2, 2021, https://www .nasa.gov/hrp/bodyinspace.

31. Ibid.

32. Ibid.

33. Ibid.

34. Harold Jackson, "John Glen obituary," *Guardian*, December 6, 2016, https://www.theguardian.com/science/2016/dec/08/john-glenn -obituary.

CHAPTER 12

1. "Train Like an Astronaut with Mike Hopkins," NASA, July 15, 2013, https://www.nasa.gov/content/train-like-an-astronaut-with-mike-hopkins.

2. NASA, "Train Like an Astronaut," YouTube video, 2:41, July 24, 2013, https://www.youtube.com/watch?v=7oBvNxbTF28.

3. Ibid.

4. NASA, "ARED Demo and Interview with Bob Tweedy," YouTube video, 7:26, February 1, 2012, https://www.youtube.com/watch?v =Uxz7u5LDRUo.

5. "International Space Station Advanced Resistive Exercise Device (ARED)," Tech Briefs, July 1, 2021, https://www.techbriefs.com /component/content/article/tb/pub/briefs/bio-medical/39434.

6. "Apollo Inflight Exerciser," Smithsonian National Air and Space Museum, January 7, 2016, https://airandspace.si.edu/stories/editorial /apollo-inflight-exerciser.

7. Ibid.

8. Ibid.

9. "Exercise Device for Orion to Pack Powerful Punch," NASA, July 14, 2016, https://www.nasa.gov/feature/exercise-device-for-orion-to-pack -powerful-punch.

10. Ibid.

11. "New Danish Space Exercise Machine Completes Testing at NASA," Danish Aerospace, April 25, 2019, https://www.danishaerospace.com/en /news/new-danish-space-exercise-machine-completes-testing-at-nasa.

12. Gary Jordan, "Artificial Gravity," March 26, 2021, interviews Bill Paloski in *Houston We Have a Podcast*, episode 188, MP3 audio, 46:48, https://www.nasa.gov/johnson/HWHAP/artificial-gravity.

13. "Wheels in the Sky," NASA Science, May 26, 2000, https://science.nasa.gov/science-news/science-at-nasa/2000/ast26may_1m.

14. Lori Ploutz-Snyder, "Integrated Resistance and Aerobic Training Study," NASA Human Research Roadmap, last published July 30, 2021, https://humanresearchroadmap.nasa.gov/tasks/task.aspx?i=573.

15. NASA Johnson, "Interview with Lori Ploutz-Snyder, PhD about SPRINT," YouTube video, 5:46, February 1, 2012, https://www.youtube.com/watch?v=X-ieJKwytgI.

16. Kirk L. English, et al., "High Intensity Training during Spaceflight: Results from the NASA Sprint Study," *npj Microgravity* 6, no. 21 (2020), https://www.nature.com/articles/s41526-020-00111-x.

17. Ingrid Skjong, "How to Get the Best Workout at Home," *New York Times*, February 2, 2021, https://www.nytimes.com/2021/02/02/realestate/how-to-get-the-best-workout-at-home.html.

18. Andrew Wagner, "20 Years on Station Lead to Advances on Earth," NASA Spinoff, last updated Nov 3, 2020, https://www.nasa.gov/directorates/spacetech/spinoff/20_years_on_station_lead_to_advances_on_earth.

19. "Bowflex System Spurs Revolution in Home Fitness," NASA Spinoffs, 2019, https://spinoff.nasa.gov/Spinoff2019/cg_1.html.

20. OYO Fitness, "OYO Personal Gym — Total Body Package," n.d., https://www.oyofitness.com/product/oyo-personal-gym-total-body/.

21. Benji Jones, "For Birds, a Steady Head Is the Key to Incredible Focus," National Audubon Society, February 6, 2018, https://www.audubon.org/news/for-birds-steady-head-key-incredible-focus.

22. Johns Hopkins Medicine, "Giving Gaze Instability a New Look," *HeadLines*, John Hopkins Medicine, May 31, 2013, https://www.hopkinsmedicine.org/news/publications/headlines/headlines_summer_2013/giving_gaze_instability_a_new_look_.

23. Jacob J. Bloomberg, "The Effects of Prolonged Space Flight on Head and Gaze Stability During Locomotion (DSO 614)," Life Sciences Data Archive, n.d., https://lsda.jsc.nasa.gov/Experiment/exper/857.

24. "Astronaut William Thornton, Who Invented Shuttle Treadmill, Dies at 91," collectSPACE.com, January 14, 2021, http://www.collectspace.com/news/news-011421a-astronaut-william-thornton-obituary.html.

1. Sarah Hampson, "Rick Hansen: Usually the Biggest Demon Is What's Inside Your Head," *Globe and Mail*, March 19, 2010, https://www .theglobeandmail.com/life/rick-hansen-usually-the-biggest-demon-is -whats-inside-your-head/article4310438/.

2. Jay C. Buckey, "Countermeasures for Space-related Bone Loss," in *Space Physiology* (New York: Oxford University Press, 2006), 14.

3. NIH Osteoporosis and Related Bone Disease National Resource Center, "What Is Bone?" October 2018, https://www.bones.nih.gov /health-info/bone/bone-health/what-is-bone.

4. Ibid.

5. Oddom Demontiero, Christopher Vidal and Gustavo Duque, "Aging and Bone Loss: New Insights for the Clinician," *Therapeutic Advances Musculoskeletal Disease* 4, no. 2 (2012): 61–76.

6. Ali H. Otom and M. Rami Al-Ahmar, "Bone Loss Following Spinal Cord Injury," *Journal of Restoratology* 2 (May 23, 2014): 81–84, https:// www.dovepress.com/bone-loss-following-spinal-cord-injury-peer -reviewed-fulltext-article-JN.

7. Ibid.

8. Waldo Cabrera, "Two Ladies Discuss What It's Like Living With Osteogenesis Imperfecta," My Long Island TV, MyLITV.com, YouTube video, 3:40, November 2, 2014, https://www.youtube.com/watch?v =RBMO83Lir34.

9. Francis H. Glorieux, "A Disease of the Osteoblast," *The Lancet Supplement* 358 (December 2001), s45. https://www.thelancet.com /pdfs/journals/lancet/PIIS0140673601070581.pdf.

10. "Bisphosphonates," Osteoporosis Canada, https://osteoporosis.ca /table-of-contents/bisphosphonates/.

11. Maureen C. Ashe, et al., "Prevention and Treatment of Bone Loss after a Spinal Cord Injury: A Systematic Review," *Topics in Spinal Cord Injury Rehabilitation* 13, no. 10 (2007): 123–145, https://www.ncbi.nlm .nih.gov/pmc/articles/PMC3389041/.

12. Buckey, 12 (see chap. 13, n. 2).

13. Andrea Wysocki, et al., "Whole-Body Vibration Therapy for Osteoporosis:

State of the Science," *Annals of Internal Medicine*, 155, no. 10 (November 2011): 1.

14. Beck, et al., "Vibration Therapy to Prevent Bone Loss and Falls Mechanisms and Efficacy," *Current Osteoporosis Reports* 13, no. 6 (December 2015): 381–9.

15. Buckey, 25 (see chap. 13, n. 2).

16. Vinita Marwhaha Madill, "A Space Suit that Squeezes," *American Scientist*, August 18, 2015. https://www.americanscientist.org/blog/the -long-view/a-space-suit-that-squeezes

17. Ibid.

18. James M. Waldie and Dava J. Newman, "A Gravity Loading Countermeasure Skinsuit" *Acta Astronautica* 68 (2011):722–730, http:// web.mit.edu/aeroastro/www/people/dnewman/pdf2/WaldieNewman _AA_GLCS_2011.pdf.

19. Ibid.

20. Buckey, 24 (see chap. 13, n. 2).

21. Mariya Stavnichuk, et al., "A Systematic Review and Meta Analysis of Bone Loss in Space Travelers," *npj Microgravity* 6, no. 13, May 5, 2020, https://doi.org/10.1038/s41526-020-0103-2.

22. Buckey, 23 (see chap. 13, n. 2).

23. Buckey, 25 (see chap. 13, n. 2).

24. Toshio Matsumoto, "Bisphosphonates as a Countermeasure to Spaceflight-Induced Bone Loss," JAXA Human Spaceflight Technology Directorate, February 26, 2021, https://humans-in-space .jaxa.jp/en/biz-lab/experiment/theme/detail/000918.html.

25. Buckey, 21 (see chap. 13, n. 2).

26. Brian Dunbar, "Human Performance Centrifuge (1.98-Meter Radius Centrifuge), NASA, n.d., https://www.nasa.gov/ames/research/space -biosciences/human-performance-centrifuge.

27. Mary Robinette Kowal, "Wally Funk is Defying Gravity and 60 Years of Exclusion From Space," *New York Times*, July 19, 2021, https://www .nytimes.com/2021/07/19/science/wally-funk-jeff-bezos.html.

28. Paul Dybala, "Interview with Senator John Glenn, Astronaut & US Senator," Audiology Online, June 12, 2006. https://www.audiologyonline .com/interviews/interview-with-senator-john-glenn-1516.

29. "STS-95," Mission Archives, NASA, John F. Kennedy Space Center,

n.d., https://www.nasa.gov/mission_pages/shuttle/shuttlemissions
/archives/sts-95.html.

30. Tamar Tchkonia, Allyson K. Palmer, James L. Kirkland, "New Horizons:
Novel Approaches to Enhance Healthspan Through Targeting Cellular
Senescence and Related Aging Mechanisms," *Journal of Clinical
Endocrinology & Metabolism* 106, no. 3 (March 2021): e1481–e1487.
https://doi.org/10.1210/clinem/dgaa728.

31. National Institute on Aging, "Does Cellular Senescence Hold Secrets
for Healthier Aging?" July 13, 2021, https://www.nia.nih.gov/news/does
-cellular-senescence-hold-secrets-healthier-aging.

CHAPTER 14

1. Steve Price, Tony Phillips and Gil Knier, "Staying Cool on the ISS,"
NASA, March 20, 2001, https://science.nasa.gov/science-news/science
-at-nasa/2001/ast21mar_1.

2. PA Media, "It's Ok, You Are Not Going to Fall: Spacewalk Astronaut
Reveals Heights Fear," *Times of Malta*, May 4, 2018, https://timesofmalta
.com/articles/view/its-ok-you-are-not-going-to-fall-spacewalk-astronaut
-reveals-heights.678087.

3. Helen Lane, et al., "Astronaut Health and Performance," in *Wings in
Orbit, Scientific and Engineering Legacies of the Space Shuttle 1971–2010*
(Houston, TX: NASA, 2011), 373, https://www.nasa.gov/centers/johnson
/pdf/584739main_Wings-ch5d-pgs370-407.pdf.

4. Louis Jolyon West, "Illusion," *Encyclopedia Britannica*, June 19, 2017,
https://www.britannica.com/topic/illusion.

5. "Ian P. Howard," Wikipedia, n.d., https://en.wikipedia.org/wiki/Ian
_P._Howard.

6. George Bush, "Presidential Proclamation 6158," Project on the Decade
of the Brain, Library of Congress, July 17, 1990, https://www.loc.gov
/loc/brain/proclaim.html.

7. Charles M. Oman, et al., "The Role of Visual Cues in Microgravity
Spatial Orientattion," in *The Neurolab Spacelab Mission: Neuroscience
Research in Space* (Houston, TX: NASA, 2003), 72. https://ntrs.nasa
.gov/api/citations/20030068201/downloads/20030068201.pdf.

8. David R. Williams, personal anecdote, STS-90 Mission Debrief, 1998.
9. Daven Hiskey, "Humans Have a Lot More Than Five Senses — Here Are 18," Considerable, September 12, 2019. https://www.considerable.com/health/healthy-living/humans-five-senses/.
10. Charles Q. Choi, "Your Perception of Gravity Is All Relative, Study Finds," Live Science, April 27, 2011, https://www.nbcnews.com/id/wbna42788105.

CHAPTER 15

1. "Emergence (episode)," Memory Alpha Fandom, Wiki, n.d., https://memory-alpha.fandom.com/wiki/Emergence_(episode).
2. Chris Farnell, "How to Transform Your Smartphone into a Real World *Star Trek* Tricorder," PCWorld, March 24, 2017, https://www.pcworld.com/article/3181416/how-to-transform-your-smartphone-into-a-real-world-star-trek-tricorder.html.
3. Elizabeth Howell, "Live Long and Prosper: '*Star Trek*'-Like 'Tricorder' Wins $2.6M Prize," Space.com, April 13, 2017, https://www.space.com/36461-star-trek-like-tricorder-medical-device.html.
4. "DxtER™: A New Kind of Medical Device," Basil Leaf Technologies, http://www.basilleaftech.com/dxter.
5. Chris Foresman, "How *Star Trek* Artists Imagined the iPad . . . Nearly 30 Years Ago," Ars Technica, September 10, 2016, https://arstechnica.com/gadgets/2016/09/how-star-trek-artists-imagined-the-ipad-23-years-ago/.
6. Yahel Galili, "What Gaming Has Taught Me About User Onboarding," UX Planet, November 12, 2020, https://uxplanet.org/what-gaming-has-taught-me-about-user-onboarding-42fdf7f55bc9.
7. Mihaly Csikszentmihalyi, "Flow, the Secret to Happiness," TED, 2004, 18:42, https://www.ted.com/talks/mihaly_csikszentmihalyi_flow_the_secret_to_happiness/transcript?language=en.
8. Lazaros Michailidis, et al., "Flow and Immersion in Video Games: The Aftermath of a Conceptual Challenge," *Frontiers in Psychology*, September 5, 2018, https://doi.org/10.3389/fpsyg.2018.01682.
9. "Addictive Behaviours: Gaming Disorder," World Health Organization,

October 22, 2020, https://www.who.int/news-room/q-a-detail/addictive-behaviours-gaming-disorder.

10. Ibid.

11. Jill A. Chafin, "3 Awesome Apps to Help Gamify Family Chores," LifeSavvy, February 18, 2020, https://www.lifesavvy.com/12266/3-awesome-apps-to-help-gamify-family-chores/.

12. Jeanne Meister, "Gamification In Leadership Development: How Companies Use Gaming To Build Their Leader Pipeline," *Forbes*, September 30, 2013, https://www.forbes.com/sites/jeannemeister/2013/09/30/gamification-in-leadership-development-how-companies-use-gaming-to-build-their-leader-pipeline/.

13. Christo Dichev, et al., "Gamifying Educations: What Is Known, What Is Believed And What Remains Uncertain: A Critical Review," *International Journal of Educational Technology in Higher Education* 14, no. 9, February 20, 2017, https://educationaltechnologyjournal.springeropen.com/articles/10.1186/s41239-017-0042-5.

14. "All Badges," Khan Academy, https://www.khanacademy.org/badges.

15. Karl Kapp, "Two Types of #Gamification," Karl M. Kapp, March 25, 2013, http://karlkapp.com/two-types-of-gamification/.

16. Gary Jordan, "The Overview Effect," August 30, 2019, interviews Frank White in *Houston We Have a Podcast*, episode 107, transcript, last updated September 4, 2019, https://www.nasa.gov/johnson/HWHAP/the-overview-effect/.

17. "Astronaut Quotes," Overview Institute, August 16, 2021, https://web.archive.org/web/20210816050342/https://overviewinstitute.org/astronaut-quotes/.

18. Ibid.

19. Liz Austin Peterson, "Astronaut Mourns His Mother from Orbit," NBC News, December 20, 2017, https://www.nbcnews.com/id/wbna22348532.

20. "NASA Remembers Sept. 11," NASA, September 11, 2021, https://www.nasa.gov/topics/nasalife/features/sept11.html.

21. Peterson, "Astronaut mourns his mother" (see chap. 15, n. 19).

22. David Dezell Turner, "Headspace: How Space Travel Affects Astronaut Mental Health," *Angles*, MIT Comparative Media Studies/Writing Program, 2019, https://cmsw.mit.edu/angles/2019/headspace-how-space-travel-affects-astronaut-mental-health/.

23. Sergio Arangio, "Cochrane Company Making Virtual Reality for Astronauts," CTV, Northern Ontario, last updated Sunday March 7, 2021, https://northernontario.ctvnews.ca/cochrane-company-making -virtual-reality-for-astronauts-1.5337227.

24. Abe Megahed, "Just in Time Simulation Platform," NASA, Human Research Roadmap, last published July 7, 2021, https:// humanresearchroadmap.nasa.gov/tasks/task.aspx?i=595.

25. "NASA, Microsoft Collaborate to Bring Science Fiction to Science Fact," NASA, June 25, 2015, https://www.nasa.gov/press-release/nasa -microsoft-collaborate-to-bring-science-fiction-to-science-fact.

26. Thomas Ormston, "Time Delay Between Mars and Earth," *Mars Express* (blog), August 5, 2012, https://blogs.esa.int/mex/2012/08/05 /time-delay-between-mars-and-earth/.

27. German Lopez, "*Pokémon Go*, Explained," Vox.com, updated August 5, 2016, https://www.vox.com/2016/7/11/12129162/pokemon-go-android-ios -game.

28. Andrea Peterson, "Holocaust Museum to Visitors: Please Stop Catching Pokémon Here," *Washington Post, The Switch*, July 12, 2016, https://www.washingtonpost.com/news/the-switch/wp/2016/07/12/ holocaust-museum-to-visitors-please-stop-catching-pokemon-here/.

29. "*Pokémon Go* Sparks Controversy Over Privacy & Security," New Net Technologies (blog archive), n.d., https://www.newnettechnologies.com /pokemon-go-sparks-controversy-over-privacy-security.html.

30. "U.S. Navy and NASA Collaborate on Augmented Reality Displays," *The Maritime Executive*, July 17, 2019, https://www.maritime-executive .com/article/u-s-navy-and-nasa-collaborate-on-augmented-reality -displays.

31. "Wireless Body Sensors for Medical Care in Space," CSA, December 23, 2015, https://www.asc-csa.gc.ca/eng/blog/2015/12/23/wireless-body -sensors-for-medical-care-in-space.asp.

32. Prajjwal Maurya, "Artificial Intelligence in Space Exploration," Analytics Vidhya, January 13, 2021, http://analyticsvidhya.com/blog/2021/01 /artificial-intelligence-in-space-exploration/.

33. Eric Ravenscraft, "What Is the Metaverse, Exactly?" *WIRED*, November 25, 2021, https://www.wired.com/story/what-is-the-metaverse/.

34. David Mikkelson, "Did the 'Father of Radio' Say Man Would Never

Reach the Moon?" Snopes, August 30, 2019, https://www.snopes.com
/fact-check/de-forest-moon-flight/.

35. John F. Kennedy Presidential Library and Museum, "Address to Joint
Session of Congress May 25, 1961," n.d., https://www.jfklibrary.org/learn
/about-jfk/historic-speeches/address-to-joint-session-of-congress-may-25
-1961.

CHAPTER 16

1. Dean Praetorius, "Bill Gates Didn't Get Why Gmail Had No Size
Limit," *Huffington Post*, April 14, 2011, https://www.businessinsider.com
.au/bill-gates-didnt-get-why-gmail-had-no-size-limit-2011-4.

2. David Pogue, "Use It Better: The Worst Tech Predictions of All Time,"
Scientific American, January 18, 2012, https://www.scientificamerican
.com/article/pogue-all-time-worst-tech-predictions/.

3. David Milstead, "Even Geniuses Make Mistakes," *New Scientist*,
August 18, 1995, https://www.newscientist.com/article/mg14719916-000
-even-geniuses-make-mistakes/.

4. Kiona N. Smith, "The Correction Heard 'Round The World: When The
New York Times Apologized to Robert Goddard," *Forbes*, July 19, 2018,
https://www.forbes.com/sites/kionasmith/2018/07/19/the-correction
-heard-round-the-world-when-the-new-york-times-apologized-to-robert
-goddard/?sh=17a95a2b4543.

5. Ibid.

6. Charles W. Lloyd, et al., "Space Radiation," NASA Human Research
Program, n.d., https://www.nasa.gov/sites/default/files/atoms/files/space
_radiation_ebook.pdf, 23.

7. Clare Skelly, "Nuclear Propulsion Could Help Get Humans to Mars
Faster," NASA, February 12, 2021, https://www.nasa.gov/directorates
/spacetech/nuclear-propulsion-could-help-get-humans-to-mars-faster.

8. Ibid.

9. National Academies of Sciences, Engineering, and Medicine, *Space
Nuclear Propulsion for Human Mars Exploration* (Washington, DC:
The National Academies Press, 2021), https://doi.org/10.17226/25977.

10. Eric Berger, "Report: NASA's Only Realistic Path for Humans on

Mars Is Nuclear Propulsion," Ars Technica, February 12, 2021, https://arstechnica.com/science/2021/02/report-nasas-only-realistic-path-for-humans-on-mars-is-nuclear-propulsion/.

11. Ibid.

12. Daniel Oberhaus, "NASA's Mars Rover Will Be Powered by US-Made Plutonium," *WIRED*, last updated July 30, 2020, https://www.wired.com/story/nasas-mars-rover-will-be-powered-by-us-made-plutonium/.

13. "NASA Outlines Lunar Surface Sustainability Concept," NASA, last updated April 3, 2020, https://www.nasa.gov/feature/nasa-outlines-lunar-surface-sustainability-concept.

14. E.L. Scheller, et al., "Long-Term Drying of Mars by Sequestration of Ocean Scale Volumes of Water in the Crust," *Science* 372, no. 6537: 52–62, March 16, 2021, https://science.sciencemag.org/content/372/6537/56.

15. Jim Wilson, et al., "NASA's MAVEN Reveals Most of Mars' Atmosphere Was Lost to Space," NASA, last updated August 6, 2017, https://www.nasa.gov/press-release/nasas-maven-reveals-most-of-mars-atmosphere-was-lost-to-space.

16. Joey Roulette, "Mars Might Be Hiding Most of Its Old Water Underground, Scientists Say," *The Verge*, March 16, 2021, https://www.theverge.com/2021/3/16/22332649/water-on-mars-martian-researchers-nasa.

17. Andrew Good, et al., "NASA InSight's 'Mole' Ends Its Journey on Mars," NASA, last updated Jan 14, 2021, https://www.nasa.gov/feature/jpl/nasa-insight-s-mole-ends-its-journey-on-mars.

18. "What Is NASA's Asteroid Redirect Mission?" NASA, last modified August 13, 2018, https://www.nasa.gov/content/what-is-nasa-s-asteroid-redirect-mission.

19. Evan Ackerman, "NASA Study Proposes Airships, Cloud Cities for Venus Exploration," *IEEE Spectrum*, September 16, 2020, https://spectrum.ieee.org/aerospace/space-flight/nasa-study-proposes-airships-cloud-cities-for-venus-exploration.

20. Adam Mann, "Why Wait for NASA? Let's Start Planning a Manned Mission to Europa Now," *WIRED*, September 28, 2013, https://www.wired.com/2013/09/objective-europa/.

21. James Miller, "How Long Would a Spacecraft Take to Reach Proxima

Centauri," Astronomy Trek, September 24, 2016, https://www
.astronomytrek.com/how-long-would-a-spacecraft-take-to-reach
-proxima-centauri/.

22. Mario Borunda, "Warp Drives: Physicists Give Chances of Faster-
Than-Light Space Travel a Boost," *The Conversation*, April 23, 2021,
https://theconversation.com/warp-drives-physicists-give-chances-of
-faster-than-light-space-travel-a-boost-157391.

23. Ibid.

24. Oliver James, et al., "Visualizing Interstellar's Wormhole," *American
Journal of Physics* 83, no. 6 (2015): 486–499, last revised April 22, 2015,
https://arxiv.org/abs/1502.03809.

25. Mike Wall, "The Science of 'Interstellar': Black Holes, Wormholes and
Space Travel," Space.com, November 10, 2014, https://www.space.com
/27701-interstellar-movie-science-black-holes.html.

26. Lee Billings, "Parsing the Science of Interstellar with Physicist Kip
Thorne," *Scientific American*, November 28, 2014, https://blogs
.scientificamerican.com/observations/parsing-the-science-of-interstellar
-with-physicist-kip-thorne/.

27. "Neptune Approach," Jet Propulsion Laboratory, California Institute of
Technology, NASA, n.d., https://voyager.jpl.nasa.gov/mission/science
/neptune/#:~:text=Voyager%202%20traveled%2012%20years,from%20
June%20to%20October%201989.

28. Joachim Boaz, "List of Generation Ship Science Fiction Novels/Short
Stories," Science Fiction and Other Suspect Ruminations, https://
sciencefictionruminations.com/sci-fi-article-index/list-of-generation
-ship-novels-and-short-stories/.

29. "Fantastic Voyage," IMDb, n.d., https://www.imdb.com/title
/tt0060397/.

INDEX

oldest people on flights, 108, 117

Oman, Chuck, 143

Onboard Short-Term Plan Viewer, 50, 51

onychoschizia, onychorrhexis, and onychogryphosis, 79

opposition class missions, 59–60

orbital debris, 75

orientation
and gravity, 139–42, 143
problems in, 24–30
and senses, 143–44

osteoblast, 131

osteogenesis imperfecta (brittle bone disease), 131–32

osteoporosis (bone loss), 130, 132–33, 134–35, 136

"overview effect," 150–51

OYO Personal Gym, 126

oyster toadfish, 29

packages to astronauts, 54–55

Paloski, Bill, 123

Parazynski, Scott, 108

Parmitano, Luca, 75

passing waste, 33–34, 36, 37

pathophysiology, xiii

Peake, Tim, 2, 27

Peregrine Mission, 163

Perseverance rover, 88, 162

personal support team, 11, 12

Pettit, Don, 83

physical activity. *See* exercise

physician astronaut, xii–xiii, xiv

physicians on Earth, xiii

physics, and time, 56–57

Pingvin (or Penguin) suit, 133–34

Pinocchio illusion, 101

pitch, 25

planetary protection (from infectious disease)
anti-contamination measures, 88, 89–90
as concept, 86–87
description, 85
and Moon, 92
NASA's work, 88–90, 94–95
protocols, 94–95, 97
requirements for, 97

plant growing, 41, 42

Ploutz-Snyder, Lori, 124

plutonium, 162

Pogue, William, 47

Pokémon Go, 154–55

potassium supplements, 39–40

pre-breathe, 73–74

preconditioning ("prehab"), 8

predictions about the future, 159–60

pressure, 69

private spaceflight, 108, 160

Project EDEN, 152–53

proprioception
description and examples, 99–101
disorders, 103–4
experiments, 101–3

proteobacteria, 84

psychologists, 11

"puffy face–bird leg" syndrome, 15–16, 20

pulsed wave Doppler ultrasound, 18

This book is also available as a Global Certified Accessible™ (GCA) ebook. ECW Press's ebooks are screen reader friendly and are built to meet the needs of those who are unable to read standard print due to blindness, low vision, dyslexia, or a physical disability.

At ECW Press, we want you to enjoy our books in whatever format you like. If you've bought a print copy just send an email to ebook@ecwpress.com and include:

- the book title
- the name of the store where you purchased it
- a screenshot or picture of your order/receipt number and your name
- your preference of file type: PDF (for desktop reading), ePub (for a phone/tablet, Kobo, or Nook), mobi (for Kindle)

A real person will respond to your email with your ebook attached. Please note this offer is only for copies bought for personal use and does not apply to school or library copies.

Thank you for supporting an independently owned Canadian publisher with your purchase!

This book is made of paper from well-managed FSC® - certified forests, recycled materials, and other controlled sources.